David de Holanda Campelo

Physiology of tree species as a function of water supply

David de Holanda Campelo

Physiology of tree species as a function of water supply

An analysis of the physiological parameters of forest species under different water regimes

ScienciaScripts

Imprint

Cover image: www.ingimage.com

This book is a translation from the original published under ISBN 978-3-330-75787-5.

Publisher:
Sciencia Scripts
is a trademark of
Dodo Books Indian Ocean Ltd. and OmniScriptum S.R.L publishing group

120 High Road, East Finchley, London, N2 9ED, United Kingdom
Str. Armeneasca 28/1, office 1, Chisinau MD-2012, Republic of Moldova, Europe
Printed at: see last page
ISBN: 978-620-8-29252-2

SUMMARY

SUMMARY

The forestry sector has grown in recent years, driven by its growing share in national exports. The demand for new areas grows every year, and in some micro-regions the timber market is of local importance, such as the Marco furniture hub in the state of Ceará. However, the main obstacle to planting tree species in this region is its low water availability, which is one of the main issues discussed in relation to the impacts of climate change, especially in arid and semi-arid regions. In order to assess the effect of water stress on the development of tree plants, an experiment was conducted with six forest species: *Astronium fraxinifolium* (Gonçalo Alves), *Calophyllum brasiliense* Cambess. (Guanandi), *Tabebuia serratifolia* (Vahl.) Nich. (Ipê-amarelo), *Tabebuia impetiginosa* (Ipê-rosa), *Simarouba amara* Aubl. (Marupà) and *Swietenia macrophylla* King. (Mogno) under irrigated and rainfed conditions. The experiment was conducted in the municipality of Acaraù, Cearâ, Brazil (03°27'06"S; 40°08'48"W, 20m). Soil moisture, gas exchange, water use efficiency, leaf temperature, photosynthetic efficiency in the use of nitrogen and phosphorus, SPAD index, chlorophyll fluorescence, specific leaf area, degree of leaf succulence, plant height, stem diameter and absolute and relative growth rates were evaluated. The water deficit caused by seasonal rainfall induces reductions in the rate of photosynthesis, stomatal conductance and transpiration. Plants under conditions of water restriction increase the intrinsic and momentary efficiency of water use in the drier periods. In dry periods, irrigated plants control leaf temperature better, keeping it lower than the air temperature. The internal/external CO_2 ratio was negatively affected in mahogany, guanandi and ipê-amarelo during the driest periods. The photosynthetic efficiency of nitrogen and phosphorus use is reduced by water deficit. Low water availability in the soil reduces the quantum efficiency of photosystem II in mahogany and guanandi. The SPAD index is reduced in the drier season in mahogany, guanandi and ipê-amarelo. The species ipê-amarelo, ipê-rosa and marupà irrigated for up to 1 year maintained a higher pattern of height and stem diameter than the permanently irrigated plants over the 24-month period. Water stress caused a drop in absolute and relative growth rates in height and diameter in the six species studied. Marupa and ipê-amarelo are less sensitive to the effect of water deficit, while mahogany, guanandi and gonçalo alves are more affected.

Keywords: Water stress. Tree plants. Water in the soil.

1 INTRODUCTION

The planted forest products sector in Brazil is currently increasing its share of exports and is very important in terms of job creation. According to data from the Brazilian Association of Planted Forest Producers, exports in 2011 amounted to US$ 7.97 billion, which corresponds to 3.1% of Brazil's total. Imports totaled US$ 2.0 billion, giving a total trade balance of US$ 5.73 billion, 19.2% of Brazil's total (ABRAF, 2012). In the world, the total area planted is approximately 50 million hectares in tropical regions, with a rate of new plantings of around 3 million hectares per year (STAPE *et al.*, 2004).

In Brazil, the main sector is pulp production, which accounts for around 37.5% of all wood produced, with the eucalyptus and pine species accounting for the largest areas of planted forests. This is followed by the production of sawn timber with 15.8%, panels with 7.8% and plywood with 3.5% of wood consumption. The production of firewood, charcoal and other forestry products account for the remaining 35.4% (ABRAF, 2012). In terms of planted area, the states of Minas Gerais (22.9%), Paranà (14.9%), São Paulo (14.3%) and Bahia (13.3%) stand out as the largest producers and account for more than half of Brazil's wood production (ABRAF, 2012).

The great growth trend of the forestry sector in Brazil has led to discussions about better ways of making the productivity of forest species viable, which shows that greater investment in research in this sector is necessary. Despite the high productivity of species such as eucalyptus and pine in most regions, there is still a lack of research into how to enable the production of other species with timber potential in regions that represent new frontiers for the expansion of the forestry sector, such as the Northeast.

There are a large number of species with timber potential that have yet to be studied (ZENID, 2000; IPEF, 2013) and this great diversity of promising species requires better research into their efficiency in producing wood for the different forestry sectors. Factors such as adaptation to different types of climate and soil are still the subject of scarce and in-depth studies for most forest species, especially in regions where forestry is of local importance, such as the lower Acaraù region in the state of Ceará. There is little information on native species that are known to be adapted to the climate of this region and that meet its needs for economic purposes. The high water deficit and poor distribution of rainfall, as well as the shallow soils with low water retention capacity and low levels of nutrients and organic matter, are some of the aspects that need to be researched.

Among the important aspects to be studied is water deficiency, considering that the region's climate has an average annual rainfall of 900 mm and average annual evaporation of 1,600 mm and is characterized by irregular rainfall (DNOCS, 2013). In these conditions, evaporation exceeds precipitation, resulting in a water deficit for most of the year.

Water deficiency is a limiting factor in plant development, affecting various metabolic processes in plants such as cell expansion, stomatal closure, reduced photosynthesis and transpiration, directly affecting the vegetative development of plants.

Knowledge of ecophysiological aspects is of great importance in order to assess the extent to which species are influenced by water stress, climate and soil conditions, which has a direct impact on plant growth and development. Therefore, more in-depth physiological studies of species subjected to the climate of the semi-arid region, especially looking at the effects of water stress, is of great interest in order to improve studies on tree species for timber production in the Northeast.

The general objective of this study is to evaluate the effect of water stress on the development of the species Gonçalo Alves (*Astronium fraxinifolium*); Guanandi (*Calophyllum brasiliense* Cambess.); Ipê -Amarelo (*Tabebuia serratifolia* (Vahl.) Nich.); Pink Ipe (*Tabebuia impetiginosa*); Marupà (*Simarouba amara* Aubl.) and Mahogany (*Swietenia macrophylla* King), under rainfed and irrigated conditions in the Baixo Acaraù Irrigated Perimeter, Cearà, by means of physiological responses and growth measurements.

2. LITERATURE REVIEW

2.1. Economic aspects of the forestry sector

2.1.1. Forestry sector in Brazil

The forestry sector can be seen as a strategic activity for Brazil's development, both for agro-industry and for the energy sector, with the potential to supply the domestic market and export. This great potential is due to favorable climatic conditions and the wide availability of areas for the activity. In short, Brazil could expand its planted forest area from the current 7 million ha (eucalyptus, pine and other planted species) to around 15 to 16 million in 10 years, which would require investments of around R$ 40 billion (or approximately US$ 20 billion) and would generate around 200,000 jobs in rural areas. At the same time, it will be necessary to develop the various segments of the wood-consuming industry (industrial logs, sawn timber, wood panels, pulp, paper and wood bioenergy), which could represent investments of around US$ 80 billion by 2020 and generate a further 800,000 jobs in urban and rural areas (ABRAF, 2012).

The mechanically processed wood (except furniture), paper and cellulose segments as a whole exported USD 7.9 billion, an increase of 5.2% on the USD 7.5 billion exported the previous year. Similarly, the forestry sector's share of the national trade balance was also significant, accounting for 19.2% of the total balance. Despite the excellent economic results in recent years, the national forest-based industry has faced systemic competitiveness problems, such as the overvalued exchange rate, a high tax burden and infrastructure problems (ABRAF, 2012). The major obstacle to the development of this sector in Brazil today is the lack of adequate institutions to organize and develop the sector's activities with economic efficiency, legal certainty and respect for the environment.

Excluding the pulp segment, which is highly productive, it can be said that the country is far from taking advantage of its full potential in terms of the forest-based industry, both in terms of the domestic market and in terms of generating foreign currency through the export of different forest-based products. Brazil has recently entered the international wood pellet trade, exporting the product to Europe, a market that is growing rapidly (20% per year). The world market for biomass fuel, however, is currently insignificant (0.2% of world consumption), but tends to gain importance if restrictions on the consumption of fossil fuels in developed countries increase (BRASIL, 2011).

As Brazil has favorable conditions for the development of planted forests, it appears to be a future power in the sector with ample capacity to expand planted areas and productivity. Investment policies in research and development, the verticalization of the sector and the qualification of the workforce employed in the activity are key factors for obtaining better productivity and consequently the evolution of the sector in the country. From an environmental point of view, planting forests in Brazil is also strategic in terms of capturing CO_2, preserving biodiversity, providing paid environmental services, generating carbon credits, protecting water sources and producing clean energy, and from a social point of view, as a generator of investment, employment and income.

2.1.2. Forestry sector in the Northeast

The Northeast region has the third largest area of planted forests, with around 13% of the national total, and is also home to a significant pulp and paper industry, which continues to expand. The expansion of planted forests is taking place especially in Bahia and, more recently, in Piaui and Maranhao. In Pernambuco, the country's main gypsum-producing state, the significant consumption of firewood from native and clandestine sources indicates the need for government intervention to encourage the planting of species suitable for energy purposes. In the production of charcoal from planted forests, the Northeast showed growth of around 9% per year between 1990 and 2008, and accounts for around 13% of the country's charcoal production (BRASIL, 2011).

The advantages of the Northeast region for forestry activity include the attractiveness of investments due to low land prices, state incentives, exports favored by the proximity of consumer markets and the highest projected growth in reforested area and installed industrial capacity. The disadvantages and challenges include the lack of qualified labor, poorly developed infrastructure, a poorly developed regional timber market and the difficulty in obtaining good forestry yields (BRASIL, 2011). Much of the northeastern territory is socio-economically and environmentally dependent on forest resources, particularly the caatinga (a typical semi-arid formation), although there are other types of vegetation in the region. Forest-based sources account for around 30 to 50% of the region's energy matrix (CAMPELLO *et al.*, 1999). Environmental authorities are concerned about the intensification of the extraction of firewood for industrial use (plaster, red ceramics, etc.) and the production of coal from biomass extracted unsustainably from the caatinga (BRASIL, 2011).

In the state of Ceará, state incentives such as the State Forestry Program, seek to develop forestry activity in the municipalities, reforestation, the installation of seed banks and nurseries, seedling production and the increase in forest replenishment in the state, stimulating the implantation of forests with a view to increasing biomass production, protecting water sources and the soil (CEARA, 2004).

The development of forestry activity in Ceará, despite state incentives, is still a major challenge given that the state has no tradition in this sector compared to other states. The wood exploitation model is still underdeveloped, with most of it used for energy production such as firewood and coal for industrial activities (plaster, red ceramics, etc.), which is often unsustainable. From an agro-industrial point of view, in the sawn wood and wood panel sector, the activity is fairly recent, but has been growing in demand mainly through the furniture poles that represent local economic importance in some municipalities of the state located mainly in the municipalities of Fortaleza, Juazeiro do Norte,

Maracanaù, Iguatu, Marco, Caucaia, Bela Cruz and Crato. Cearà is consolidating its position as the largest furniture hub in the Northeast. Its share of the regional and national markets has grown considerably in recent years, as has the number of industries specializing in furniture production (SCIPIÂO, 2004; SEBRAE, 2008).

Furniture production still uses solid wood from other regions such as Parà, Minas Gerais and Paranà, which requires greater investment in wood production in the state in order to reduce production costs for local industries and the viability of industries linked to the wood sector (SEBRAE, 2008).

2.2. Water stress in tree plants

Stress is an external factor that exerts a disadvantageous influence on the plant (TAIZ; ZEIGER, 2013), inducing responses at all levels of the organism, which can be reversible or permanent. Lechinoski *et al.* (2007) define stress as excessive pressure from an adverse factor that tends to inhibit the normal functioning of systems.

In agricultural production, the study of plant response to water stress is one of the main issues to be discussed, given its importance for increasing agricultural production, even in areas of maximum productivity where it is a limiting factor for plant production (BRUCE *et al.*, 2002).

Water stress can lead plants to an initial destabilization of functions, followed by normalization and the development of resistance. However, if tolerance limits are exceeded, and adaptive capacity is supplanted, the result is permanent damage or often death (PAHLICH, 1993). Plant responses to stress can be understood according to the degree of exposure to a given stress: adaptation, which is genetic resistance acquired through selection over many generations; acclimatization, which can be understood as increased tolerance as a result of previous exposure to stress; and tolerance, which is a plant's ability to cope with an unfavorable environment. (TAIZ; ZEIGER, 2013).

Knowledge of the physiological mechanisms of tolerance to water stress in tropical tree plants has been widely discussed, given that the current scenario of deforestation and soil degradation reflects a problem in terms of creating sustainable forest management systems (REINHARDT *et al.*, 2000). The water-plant relationship has been studied in order to understand the processes of absorption, transportation and loss of water, as well as the survival strategies of plants subjected to environments with a lack or excess of water in the soil (MEDRI, 2002), which can trigger different degrees of stress.

In general, abiotic factors directly affect the physiological activity of plants in their different growth phases. Factors such as low light, water and nutritional availability make it more difficult for plants to establish themselves in the juvenile phase. In relation to water availability, various plant metabolic processes can be influenced, such as stomatal closure leading to a reduction in photosynthesis and transpiration and in biomass allocation patterns, affecting plant growth and is one of the conditions that most limits the primary production of ecosystems and crop yields, mainly due to the restrictions it imposes on photosynthetic carbon fixation (GRISI *et. al.,* 2008; TAIZ; ZEIGER, 2013; LARCHER, 2006).

2.3. Morphophysiological effects of water stress

2.3.1. Effects on growth

The first physiological responses to water deficit involve cell contraction, loosening of the cell wall and loss of cell turgor and consequently a reduction in plant growth (NOGUEIRA *et al.*, 2004; CERNUSAK *et al.*, 2007). As the loss of turgor is the first biophysical effect of water deficiency, turgor-related activities are the most sensitive to water deficit. Water deficiency affects all aspects of plant growth and development, influencing cell elongation and resulting in a decrease in leaf area, affecting the production and translocation of photoassimilates to new growth areas, as well as a reduction in the production and allocation of dry matter, which is directly influenced by a reduction in gas exchange and carbon balance (LUDLOW; MUCHOW, 1990; LARCHER, 2006; NOGUEIRA *et al*, 1998, SILVA *et al.*, 2003; NOGUEIRA *et al.*, 2005). The process of stem growth is less studied, but is probably affected by the same forces that limit leaf growth during stress (TAIZ; ZEIGER, 2013).

The reduction in leaf area in woody plants as a response to low soil water availability is proven by several authors (CABRAL *et al.*, 2004; COSTA; MARENCO 2007; SCALON *et al.*, 2011) and is considered to be one of the main factors leading to a drop in productivity (LARCHER, 2006). Another effect observed is a reduction in the allocation of leaf and stem biomass through an increase and accumulation in the roots, which is why there is greater growth in the root zone in response to the low availability of water in the soil and, consequently, an increase in the capacity to absorb nutrients (CORREIA; NOGUEIRA, 2004).

Water stress is one of the factors that modifies physiological and morphological processes in plants by altering the water content in the cell and inhibiting various growth processes in a way that alters plant metabolism. Thus, a better understanding of these processes in plant growth is essential for the development of sustainable management systems and a better understanding of the adaptation of species to abiotic stresses.

Reduced water availability has a direct effect on the photosynthetic capacity of plants, mainly through dehydration of the photosynthetic apparatus and indirectly through stomatal closure (LAWLOR; CORNIC, 2002). Photosynthetic capacity is an intrinsic characteristic of each plant species, and gas exchange is variable and changes according to the crop's development cycle and depends on the annual and even daily course of environmental fluctuations (light, temperature, etc.) around the plant

(LARCHER, 2006).

Under conditions of low water availability in the soil, plants develop adaptations to tolerate water stress. One of these adaptation mechanisms is the reduction in photosynthetic rate through stomatal closure due to the change in water *status* in the leaf, leading to a reduction in transpiration and the possibility of death by desiccation (SILVA *et al.*, 2009).

Stomatal closure also affects the dynamics of water vapor exchange and CO_2 absorption flow, thus limiting the entry of these gases into the plant and directly affecting photosynthesis (PIMENTEL, 2004). This decrease can also be due to the loss of turgidity in the stomatal guard cells (BARUCH, 1994; SILVA *et al.*, 2001). The decrease in carbon assimilation is observed by many authors as being one of the direct responses to water stress (LIMA *et al.*, 2010; ALBUQUERQUE *et al.*, 2013; COSTA; MARRENCO 2007). Understanding the responses of stomatal closure is of great importance, as the balance between the entry of CO_2 into the leaf and the loss of water vapor to the atmosphere through transpiration is essential for maintaining photosynthetic activity and hydration of the tissues so that the plant obtains the substrate for the biochemical reactions of photosynthesis and prevents excessive loss of water and consequent dehydration of the tissues (CHAVES *et al.*, 2002).

The variables of net photosynthesis rate (A), transpiration rate (E) and stomatal conductance (g_s) are parameters considered to be physiological variables that indicate important measures of the ability of species to establish themselves in environments with limited water availability. In this sense, indicators such as water use efficiency (A/E) and intrinsic water use efficiency (A/g_s) can be used in research to select the most appropriate species for planting in certain conditions of water availability and to adopt practical improvements in irrigation management (CORDEIRO, 2012; FARIAS, 2008).

2.3.3. Effects on chlorophyll content (SPAD index)

Chlorophylls are pigments located in the thylakoid membranes of chloroplasts and are responsible for capturing the light used in photosynthesis. They are essential for converting light radiation into chemical energy in the form of ATP and NADPH (VICTÓRIO *et al.,* 2007). Chlorophyll "a" is present in all organisms that carry out photosynthesis and is considered the main pigment, with the other pigments being accessories: chlorophyll "b" is another type of pigment found in plants, green algae and some bacteria; chlorophyll "c", in phaeophytes and diatoms and chlorophyll "d", in red algae. In the membranes of the thylakoids, photosystems I and II are responsible for capturing and converting light energy (photons) into chemical energy (ATP and NADPH). The reaction center of photosystem II is formed by a protein complex together with the P680 chlorophyll molecule, which has a very strong reductant that oxidizes the water molecule into electrons, protons and oxygen (TAIZ; ZEIGER, 2013).

Chlorophyll and carotenoid levels in leaves are used to estimate the photosynthetic efficiency of plants, as they are directly linked to the absorption and transfer of light energy, which is essential for photochemical reactions (RÊGO; POSSAMAI, 2004). Thus, high chlorophyll levels in plants favor higher photosynthetic activity due to their greater capacity to capture light (PORRA *et al.*, 1989).

According to Almeida *et al.* (2004), the growth and adaptation of plants to different environments are related to leaf chlorophyll levels, among other factors. The existence of a relationship between the green color intensity index, chlorophyll content and water stress in plants has been demonstrated, due to the loss of pigments that occurs during exposure to an environmental stress, such as drought, showing it to be a visible indicator of the plant's stress, and can be used as an indicative parameter of abiotic stresses (HOEL; SOLHAUG, 1998; CARVALHO *et al.*, 2007; CODOGNOTTO *et al.*, 2002).

2.3.4. Effects on chlorophyll fluorescence

Chlorophylls have their maximum photon absorption point in the 428 to 660 nm region (chlorophyll a) and in the 452 to 641.8 nm region (chlorophyll b) corresponding to the blue and red regions respectively, in which case the green light is not absorbed but reflected and transmitted, giving most plants their green color. When chlorophyll molecules absorb light energy (photons), they change their electronic configurations, moving to a more excited state (higher and more unstable orbitals) called *singlet 1*. When the electron returns to its original orbit, it releases the energy it absorbed from the photon in the form of light. This energy can be dissipated in three ways: photochemical dissipation (Ph), fluorescence (F) and non-photochemical dissipation (D) (CAMPOSTRINE, 2001).

At room temperature (physiological temperature, 20 to 25 °C), fluorescence is emitted light and exhibits a maximum emission point in the 682 nm range and another less pronounced point at 740 nm (KRAUSE; WEIS, 1991). The three processes of light energy dissipation by chlorophyll molecules (Ph+F+D) are competitive, i.e. changes in photosynthetic rates and heat dissipation will cause complementary changes in fluorescence emission. Thus, changes in fluorescence can show the absence or presence of impairments in the photosynthetic process. This fluorescence emitted by photosystem II (PSII) can be detected with a photodetector, i.e. a photomultiplier or photodiode sensitive to wavelengths in the 680 nm region (CAMPOSTRINE, 2001).

Chlorophyll a fluorescence measurements have been an important tool because, as well as being non-destructive, they allow qualitative and quantitative analysis of the absorption and use of light energy by photosystem II and possible relationships with photosynthetic capacity (MOUGET; TREMBLIN, 2002). In this sense, this technique has been used to investigate the damage caused by various types of stress on the photosynthetic capacity of plants, and more specifically to evaluate the effects on the photosynthetic apparatus that are directly attributed to photochemical efficiency, as it is a tool capable of diagnosing the reduction in CO_2 assimilation and the destruction of the photosynthetic structure

(TORRES NETO *et al.*, 2005; CORREIA *et al.*, 2009; GONÇALVES *et al.*, 2010). This technique is a tool of great potential in studies related to the effects of environmental factors on the physiological process in plants, as it is able to indicate the absence or presence of impairment in the photosynthetic process in plants subjected to water stress (PINCELLI, 2010).

2.3.5. Effect on photosynthetic efficiency of nitrogen and phosphorus use

Mineral elements are extremely important for the efficiency of photosynthesis, as they are integral components of enzymes and pigments or direct activators of the photosynthetic process. Nitrogen, an essential component of proteins and chlorophylls, is necessary for the formation of thylakoids and enzymes, especially the enzyme ribulose 1,5 bisphosphate carboxylase/oxygenase (Rubisco), which acts in carbon fixation (LARCHER, 2006; FYLLAS *et al.*, 2009). Inorganic phosphorus regulates the Calvin cycle and the transport of metabolites and assimilated compounds. Phosphate deficiency results in an accumulation of assimilates (sucrose and starch) in the chloroplast, depressing photosynthesis even under favorable conditions (LARCHER, 2006).

Although studies have shown that the capacity and efficiency of the photosynthetic process are altered by water stress (ALBUQUERQUE *et al.*, 2013; CORDEIRO 2012) and rainfall seasonality (PALHARES *et al.*, 2010; TROVÂO *et al.*, 2007), it is not yet known how the content of mineral elements, such as nitrogen and phosphorus, can influence the photosynthetic capacity of trees subjected to water deficiency and throughout the different rainfall seasons.

3. MATERIAL AND METHODS

3.1. Location and characterization of the experimental area

The work was carried out from September 2012 to July 2013 in an Embrapa experimental area, located in the irrigated perimeter of Baixo Acaraù, 2.2 km from the right bank of the CE-178 highway, in the municipality of Acaraù, Cearà, Brazil, with coordinates of 3° 27' 06" South latitude, 40° 08' 48" West longitude and an average altitude of 60m (Figure 1). The region's climate according to the Koppen classification is Aw Tropical Rainy: average annual rainfall of 900 mm; minimum annual temperature 22.8° C; average annual temperature 28.1° C; maximum annual temperature 34.7 °C; sunshine 2,650 h/year; average annual relative humidity 70%; average wind speed 3.0m/s and average annual evaporation 1,600 mm. The relief is fairly gentle and the soils are on average deep, well-drained, of medium or medium/light texture and very permeable (DNOCS, 2013).

The rainfall data for the period from October 2010 to October 2013 is shown in Figure 2. The data was provided by a FUNCEME weather station, located about 6 km from the experimental area in the community of Lagoa do Carneiro, Acaraù, Cearâ.

Figure 1 - Location of the experimental area. Acaraù-CE, 2013.

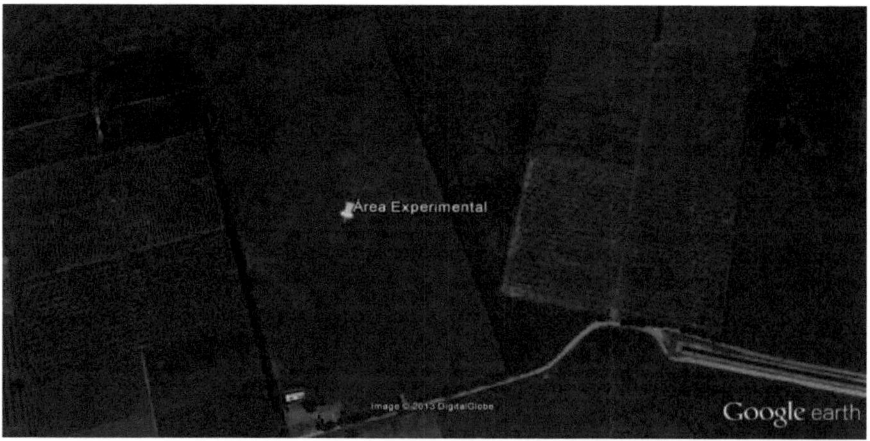

Source: (Google Earth, 2013).

Figure 2 - Monthly rainfall averages from October 2010 to October 2013, in the municipality of Acaraù - CE.

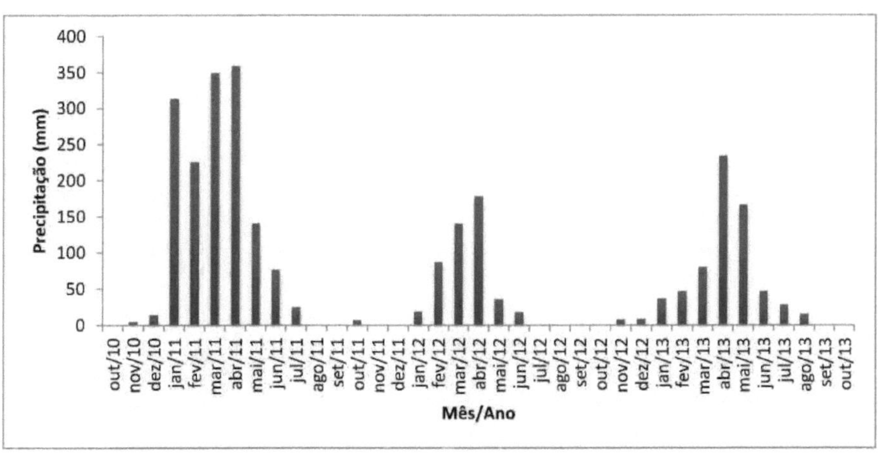

Source: Fundaçao Cearense de Meteorologia e Recursos Hidricos, 2014.

The soil in the experimental area was classified as quarzoarenic neosol (Embrapa, 2013). Soil samples were collected at depths of 0 - 20, 20 - 50 and 50 - 80 cm for chemical and physical characterization. The chemical analyses were carried out at the Soil, Water and Plants Laboratory of Embrapa Agroindustria Tropical - Fortaleza, Cearà and the physical analyses at the Soil/Water Laboratory of the Ceará Foundation for Meteorology and Water Resources (FUNCEME) - Department of Soil Sciences/UFC - Fortaleza, Cearâ. The results of the chemical and physical analyses of the soil are shown below in Tables 1 and 2.

Table 1 - Chemical characterization of the soil in the experimental area. Acaraù-CE, 2013.

Layer	pH	EC	MO	P	K^+	Mg^+	Ca^{+2}	In^{+2}	$H+Al^{+3}$	CTC	SB	V%
(cm)	Water	dS/m	g/Kg	mg/dm^3	mmolc/dm3-----							%
0 - 20	6,4	0,21	12,9	22,9	1,6	7,2	18,8	3,3	7,5	38,4	30,9	80,5
20 - 50	6,1	0,17	5,1	16,9	0,5	5,7	10,5	3,4	3,3	23,4	20,1	86,0
50 - 80	5,7	0,12	5,1	12,8	0,7	5,4	8,5	3,5	10,8	28,9	18,1	62,2

Source: Soil, Water and Plants Laboratory - Embrapa Agroindustria Tropical.

Table 2 - Physical characterization of the soil in the experimental area. Acaraù-CE, 2013.

Layer	Particle Size Composition (g/Kg)				Textural Classification
(cm)	Coarse sand	Fine Sand	Silt Clay	Natural Clay	

0 - 20	609	311	2951	23		Sand
20 - 50	602	296	3567	50		Sand
50 - 80	403	481	3185	42		White Sand

Layer	Density (g/cm3)		Moisture (g/100g)			Degree of flocculation
(cm)	Global	Particle	0.033 MPa	1.5 MPa	Useful Water	(g/100g)
0 - 20	1,51	2,69	4,02	3,11	1,91	55
20 - 50	1,50	2,71	4,38	2,95	1,43	26
50 - 80	1,50	2,72	4,62	3,27	1,35	51

Source: Fundaçao Cearense de Meteorologia e Recursos Hidricos.

3.2. Species used

3.2.1. Gonçalo Alves (Astronium fraxinifolium)

The Gonçalo-Alves is a species belonging to the *Anacardiaceae* family, with deciduous behavior. The trees can reach 25 m in height and 60 cm in DBH (diameter at breast height, measured at 1.30 m from the ground), in the Cerrado and Caatinga they commonly reach 3 to 5 m in height. It has a straight trunk; a stem of up to 8 m; bark up to 17 mm thick; alternate, imparipinnate leaves, 15 cm to 12 cm long and 2 to 3.5 cm wide. The inflorescence occurs in terminal or axillary panicles, measuring 2 to 6 cm long with white or yellow flowers with five petals and 4 mm long. The drupe-like fruit measures 1 cm in length and 3 to 5 mm in diameter and is light brown in color. The seed is oblong, 5 to 7 mm long and has a polyembryony rate of 4% (CARVALHO, 2010; SALOMÂO; ALLEM, 2001).

Figure 3 - Gonçalo Alves (*Astronium fraxinifolium*). Acaraù-CE, 2013.

Source: Author's elaboration.

It occurs naturally between latitudes 0°25'N in Amapà and 23°S in Mato Grosso do Sul, and has an attitudinal range of 15 m in Paraiba to 1,300 m in Minas Gerais. The regions of occurrence vary in average annual rainfall from 700 mm in Piaui to 2,600 mm in Amapà and average annual temperatures from 18.1 °C in Diamantina, MG to 26.5 °C in Bom Jesus do Piaui, PI. It is a characteristic calciferous species and an indicator of mesotrophic soils. It prefers well-drained sites and sandy or clay soils with high fertility and depth (CARVALHO, 2010; RATTER *et al.*, 1978; LOPEZ *et al.*, 1987; POTT; POTT 1994).

The wood is dense to very dense with an apparent specific mass (density) of 0.73 $g.cm^{-3}$ to 1.13 $g.cm^{-3}$ and a basic specific mass of 0.820 $t.m^{3}$. Heartwood is very irregular with a pinkish-beige to brown coloration; glossy surface, smooth to the touch; medium texture is uniform; irregular, diagonal or coated grain. The wood of this species is very durable, highly resistant to fungal attack and easy to work. It is used in civil and naval constructions, poles, parquets, luxury furniture, ornamental objects and turnings, as well as sheets for plywood. It provides excellent quality firewood and charcoal (BENITEZ; MONTESINOS LAGOS, 1988; BRITO; BARRICHELO, 1981; CARVALHO, 2010).

3.2.2. Guanandi (Calophyllum brasiliense Cambess.)

Guanandi belongs to the *Clusiaceae (Guttiferae)* family *and* is an evergreen tree, losing few leaves during the dry season (MARQUES, 1994). In the Northeast, it can reach heights of up to 16 m (LIMA; ROCHA, 1971), in the South, up to 25 m (REITZ *et al.*, 1978), and in the Amazon it can reach heights of up to 40 m and a DBH of 150 cm in adulthood (BASTOS, 1946). The trunk is straight and cylindrical; the stem is up to 15 m high; the bark is up to 40 mm thick; the leaves are cross-shaped (decussate), simple, elliptical, leathery, 5 to 15 cm long and 3 to 7 cm wide, with numerous, approximate, parallel lateral veins running up to the margin. The flowers are gathered in short axillary racemes or small panicles 2.5 to 6 cm long, arranged in white triflora clusters. The fruit is a fleshy, indehiscent globular drupe 19 to 30 mm long. The seeds are globular, brown in color and 14 to 22 mm in diameter (CARVALHO, 2003).

It occurs naturally between latitudes 18°N in Puerto Rico and 28°10'S in Brazil, in Santa Catarina, and has an attitudinal range of 5 m on the coast of the South, Southeast and Northeast regions, to 1,200 m in the Federal District. Outside Brazil, it reaches altitudes of up to 1,500 m (CARVALHO, 2003; JOKER; SALAZR, 2000).

The regions of occurrence have a range of average annual rainfall from 1,100 mm in the state of Sao Paulo to 3,000 mm in Parà and average annual temperatures from 18.1 °C in Diamantina, MG to 26.7 °C in Manaus, AM in alluvial soils with poor drainage, in wet, periodically flooded and marshy places, with a sandy and loamy texture, and acidic (pH 4.5 to 6.0) (CARVALHO, 2003).

The wood varies from light to moderately dense with an apparent specific mass of 0.45 to 0.65 $g.m^{-3}$; between 12% and 15% humidity and a basic specific mass of 0.49 to 0.51 $g.cm^{-3}$. The sapwood is lighter; the heartwood varies from pinkish-brown to pinkish-beige, tending towards brown; the surface is lightly polished and rough to the touch; it has a medium to coarse texture and irregular grain. The wood is considered to be durable to moderately durable; imputrescible in water and good workability, used in the manufacture of furniture and used in civil and naval constructions, barrels, joinery and carpentry in general (CARVALHO, 2003; JOKER; SALAZAR, 2000; JANKOWSKY *et al.*, 1990).

Figure 4 - Guanandi (*Calophyllum brasiliense* Cambess.). Acaraù-CE, 2013.

Source: Author's elaboration.

3.2.3. Yellow Ipe (Tabebuia serratifolia (Vahl.) Nich.)

The ipê-amarelo is a species belonging to the *Bignoniaceae* family. It is a deciduous, heliophytic plant, characteristic of dense rainforests. The trees of this species generally reach a height of 8-20 m. The trunk is 60-80 cm in diameter and the bark is 10-15 mm thick, grayish-brown in color with a fissured surface. The leaves are compound, 5-foliolate (possibly 4); the leaflets are glabrous or pubescent, 6-17 cm long and 3-7 cm wide. Inflorescence in corymb panicles with yellow flowers. The calyx and corolla of the flowers have a tubular structure with five lobes. The greenish calyx is slightly pubescent. The golden yellow corolla is 68 cm long. The fruit is a septate, leathery, glabrous, linear pod, 20-65 cm long and 2.5 to 3.5 cm thick. The numerous seeds are rectangular, laminar, light, with two short, hyaline wings (LORENZI, 2002; FERREIRA *et al.*, 2004).

Figure 5- Yellow Ipê (*Tabebuia serrati fola* (Vahl.) Nich.). Acaraù-CE, 2013.

Source: Author's elaboration.

It occurs naturally very frequently in the Amazon region and sparsely from Cearà to Sao Paulo in the Atlantic rainforest and in the semi-deciduous forest from Parà and Mato Grosso to Goiás; in the southern regions of Bahia and northern Espirito Santo it is slightly more frequent than on the rest of the coast. It is found from sea level up to an altitude of 1200m. As for soil conditions, both in the forest and in capoeira, it prefers well-drained soils located on the slopes (LORENZI, 2002; FERREIRA *et al.*, 2004).

The wood is heavy (density 1.08 g/cm^3), very hard, difficult to saw, rich in lapachol crystals, infinitely durable under any conditions, with sapwood distinct from heartwood. The surface is not very shiny, smooth, oily and has a light brown to dark brown color, with greenish reflections. The wood is suitable for heavy construction and external structures, both civil and naval, such as ship keels, bridges, dormers, poles, for floorboards, making billiard cues, walking sticks, wheel axles, etc. (LORENZI, 2002; FERREIRA *et al.*, 2004).

3.2.4. Pink Ipe (Tabebuia impetiginosa)

The Ipê-Rosa is a species belonging to the *Bignoniaceae* family and is a deciduous tree. They can reach 15 m in height and 30 cm DBH (diameter at breast height, measured at 1.30 m from the ground) in the Caatinga and up to 50 m and 100 cm DBH in the Amazon (PAULA; ALVES, 1997). The trunk is often crooked, but straight and cylindrical individuals can be found; the stem is 4 to 8 m long; the bark is up to 12 mm thick; the leaves are opposite and digitate, with petioles up to 11 cm long, usually with five leaflets with an inner margin or slightly serrated up to the apex. The flowers are pink, tubular and 4 to 7.5 cm long, gathered in a terminal panicle. The fruit is a narrow cylindrical siliqua, dehiscent, 12 to 56 cm long. The seeds are cordiform, oblong and light brown, with a membranous wing up to 3 cm long (CARVALHO, 2003; SOUZA; LIMA, 1982).

Figure 6 - Pink Ipê (*Tabebuia impetiginosa*). Acaraù-CE, 2013.

Source: Author's elaboration.

It occurs naturally between latitudes 20°N in northeastern Mexico and 28°S in northeastern Argentina. In Brazil, the southern limit of this species is 22°45'S in Paranà, and has an attitudinal range of 10 m on the coast of the Northeast to 1,400 m in the state of Sao Paulo. The regions of occurrence vary in average annual rainfall from 440 mm in Bahia to 2,500 mm in Pernambuco and average annual temperatures from 18.1 °C in Diamantina, MG to 27.2 °C in Mossor6, RN. The soils in the area of natural occurrence are sandy and moist, well-drained and with a loamy to clayey texture. Soils with low nutrient content are limiting to its growth (CARVALHO, 2003; GARRIDO, 1981).

It has dense wood with an apparent specific mass (density) of 0.92 to 1.08 g.cm-3 at 15% humidity and a basic specific mass of 0.79 $t.m^{-3}$ (MAINIERI; CHIMELO, 1989; JANKOWSKY *et al.*, 1990). Heartwood and sapwood are undifferentiated, brownish or light vanilla-brown in color, generally uniform; the surface is slightly glossy and moderately smooth to the touch; the texture is fine to medium and uniform; the grain is straight or curved. The wood is resistant to attack by xylophagous organisms, highly resistant to fungi and termites and is easy to work. It is used in construction, making sports articles, tool handles for agricultural implements, turning parts and making musical instruments, as well as providing good quality firewood (CARVALHO, 2003; PAES *et al.*, 2005; CAVALCANTE, 1982).

3.2.5. Marupà (Simarouba amara Aubl.)

Marupà belongs to the *Simaroubaceae* family *and* is an evergreen to semi-deciduous tree. They can reach 25 m in height and 80 cm in DBH (diameter at breast height, measured at 1.30 m from the ground). The trunk is straight and cylindrical in section; the stem is up to 22.5 m long; the bark is up to 20 mm thick; the compound leaves are imparipinnate with 3 to 6 pairs of 7 to 21 leaflets, 5 to 10 cm long and 3 to 3.5 cm wide. Inflorescences occur in erect, uninfloresced, highly branched terminal panicles, measuring 20 to 30 cm in length. The flowers are white or yellowish, the female flowers have 5 to 6 sepals and the male flowers are pentamerous. The fruit is a globular drupe formed from a single carpel. The seed is elliptical, measuring from 0.5 cm to 0.8 cm in length (CARVALHO, 2008; BARROSO *et al.*, 1999).

It occurs naturally between latitudes 3°N in Roraima and 21°45'S in Rio de Janeiro, and has an attitudinal range of 20 m in Parà to 1,200 m in the Federal District. The regions of occurrence vary in average annual rainfall from 760 mm in Cearà to 4,000 mm in Amapà and average annual temperatures from 20.5 °C in Guaramiranga, CE to 26.7 °C in Manaus, AM. It is a species found naturally in various types of soil (CARVALHO, 2008).

Figure 7 - Marupà (*Simarouba amara* Aubl.). Acaraù-CE, 2013.

Source: Author's elaboration.

The wood is considered light with an apparent specific mass (density) of 0.352 $g.cm^{-3}$ to 0.55 $g.cm^{-3}$ at 12% to 15% humidity (WOODCOCK, 2000). The heartwood and sapwood are indistinct and straw-white to slightly yellowish in color; the grain is straight to irregular; the texture is medium; the surface is slightly rough to the touch and not very glossy or smooth when cut. The natural durability of the wood is relatively long, it is resistant to insect attack and has good workability. It has a wide range of uses in the shoe industry, construction (internal finishes such as skirting boards, mouldings, lining boards), plywood production, furniture, the manufacture of matchsticks, crates and musical instruments. It supplies reasonably high-quality firewood (CARVALHO, 2008).

3.2.6. Mahogany (Swietenia macrophylla King)

Mahogany is a species belonging to the *Meliaceae* family. *It is* an evergreen deciduous tree and can reach up to 70 m in height and 3.5 m in DBH (diameter at breast height, measured at 1.30 m from the ground). It has a straight trunk, slightly fluted and with tabular roots at the base; a straight stem up to 27 m long; bark up to 25 mm thick; compound leaves, arranged in a spiral on the branches, paripinnate, measuring 25 to 45 cm long. They have 8 to 12 leaflets measuring 7 to 15 cm long and 3.5 to 6 cm wide. Inflorescence in dense, pyramidal axillary shoots measuring 15 to 25 cm long. Flowers of both sexes are found on the same inflorescence, with the male flowers being actinomorphic, measuring 6 to 8 mm in diameter and the female flowers similar, but with very small anthers, indehiscent and without pollen. The fruit is a woody, ovoid capsule, 10 to 22 cm long and 6 to 10 cm wide, with septate dehiscence and a brown color. The seeds are brownish-red winged and measure 8 to 25 mm in length (CARVALHO, 2006).

Figure 8 - Mahogany (*Swietenia macrophylla* King). Acaraù-CE, 2013.

Source: Author's elaboration

It occurs naturally between latitudes 20°N in Mexico (Yucatán) and 18°S in Bolivia. In Brazil, it ranges from 1°S in Maranhao to 14°S in Mato Grosso and has an attitudinal variation of 400 m in Brazil to 1,500 m in Peru (PENNINGTON, 1981). The regions of occurrence vary in average annual rainfall from 1,200 mm in Maranhao to 2,900 mm in Parà and average annual temperatures from 24.8 °C in Tarauacà, AC to 26.7 °C in Itaituba, PA (CARVALHO, 2006).

The tolerable soil conditions for mahogany vary, from poorly drained deep soils, acidic clay soils and swamps, to well-drained alkaline soils from limestone plateaus, including soils derived from igneous and metamorphic rocks, ranging from those in areas subject to periodic flooding (hydromorphic) to soils in terra firme areas (spodosols), typical of the region where they occur (CARVALHO, 2006; TEREZO, 2002).

It has moderately dense wood with an apparent specific mass (density) of 0.48 to 0.85 $g.cm^{-3}$, with 12% to 15% humidity (MAINIERI; CHIMELO, 1989). The heartwood is light brown and slightly yellowish; the sapwood is narrow and well contrasted yellowish-white; the surface is glossy, smooth to the touch, with golden reflections; the texture is medium and uniform; the grain is straight or slightly irregular (diagonal). The wood is considered to be moderately resistant to rot and highly resistant to dry wood attack. In contact with soil and humidity, it is not very durable. It is easy to work, is one of the most sought-after woods for export and is used in the country to make luxury furniture, carved doors, decorative objects, paneling, civil construction, the aviation industry, plywood, panels, laminates, high-precision scientific instruments and to make musical instruments, especially pianos (CARVALHO, 2006).

3.3. Experimental design and treatments

The experimental design adopted was that of repeated measures over time in a sub-subdivided plot layout (6 x 3 x 2) for the gas exchange variables and (6 x 2 x 2) for the photosynthetic efficiency variables in the use of P and N, SPAD index, chlorophyll fluorescence, degree of succulence and specific leaf area. For the growth variables (plant height, stem diameter, absolute and relative growth rates) the scheme adopted was (6 x 2 x 5), with the main plot consisting of six species, the sub-plot of two water regimes (irrigated and dry) and the sub-sub-plot of the evaluation times. Multiple observations were made of the same plant at the different times, considering each plant as a replication. The treatments in the experiment basically consisted of two groups: one irrigated (control treatment) and the other non-irrigated (water deficit treatment).

3.4. Setting up and condition of the experiment

3.4.1. Seedling production

The seedlings were produced using seeds collected from trees located in Ceará and Paraiba during the 2010 harvest. After collection, they were transported to the Genetic Resources and Improvement Laboratory at Embrapa Agroindustria Tropical, where they were processed and stored in a refrigerator (4 to 6 °C; 35 to 43% relative humidity). The seedlings were produced in the nursery of the Pacajus experimental field of Embrapa Agroindustria Tropical, using the direct sowing method in 288 cm tubes[3] containing a mixture: carbonized rice husk + crushed carnauba bagana + hydromorphic soil in a volumetric ratio of 3:2:2. After the end of production, which took place on average 60 days after germination, the seedlings went through a process of hardening off, which lasted another 30 days, before they were planted out in the field. The seedlings were vigorous, free of pests and diseases and without nutritional deficiencies.

3.4.2. Implementation and management

In October 2010, the species were planted in experimental plots measuring 6 x 28 m, consisting of three rows of 15 plants/row, the first and third of which were considered borders. The two plants at the ends of the central row were also excluded, giving a total of 13 plants in the useful area. The spacing used was 3 m between rows and 2 m between plants (Figure 9). The 20x20x20 cm pits received 150 g of foundation fertilizer, which consisted of 120 g of NPK 10-28-20 plus 30 g of FTE BR 12 mixed with the soil at the bottom of the pit. In addition, the soil removed when the hole was opened was returned to fill and close the hole after the seedling had been placed. Replanting was carried out up to 30 days after planting to replace dead or defective seedlings.

At the same time as planting the seedlings, Guandu Ana beans - cultivar IAPAR 43 - were sown as a form of green manure between the rows. Every month for eight months, the crops were cut back to a height of 50 cm from the ground and the plant remains were left on the soil surface. During the experimental period, at six-month intervals, each plant received training fertilizer by incorporating 50 g of NPK 10-28-20 spread in a half circle at a depth of 5 to 10 cm. At 12 months, the first stem formation pruning was carried out for species with excessive branch formation. Ant control was monitored periodically in the area, with the occasional application of granulated and powdered formicide. To control invasive plants, manual weeding was carried out in the planting rows.

Figure 9 - Sketch of the sub-plot. Acaraù-CE, 2013.

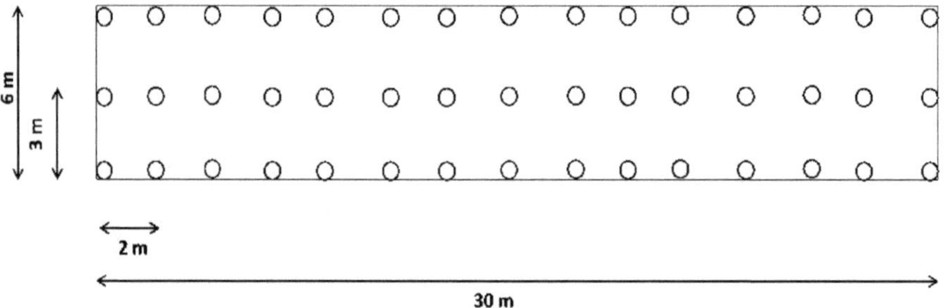

Source: Author's elaboration.

3.4.3. Irrigation management

In the first year, the entire area was irrigated and after this period it was suspended on one side until the end of the experiment, dividing it into two sub-areas (sub-plots), one which was irrigated continuously (irrigated regime) and the other which was suspended at the end of the first year of cultivation (non-irrigated regime). A micro-sprinkler irrigation system was used and irrigation was managed as follows: in the first year the irrigation shift was 1 day, irrigation time 2 hours, volume per micro-sprinkler 80 L/day, with volume per plant 16 L/day and a total applied blade of 2.7 mm/day. After the first year, the irrigation shift was 2 days, irrigation time 7.5 hours, volume per micro-sprinkler 300 L/day, with a volume per plant of 60 L/day, with an applied blade of 10 mm/day and an annual blade of 1200 mm (considering 8 months).

3.4.4. Evaluation times

For the variables of gas exchange, chlorophyll fluorescence, SPAD index and soil moisture, three evaluations were carried out on 11/22/12, 02/07/2013 and 05/17/2013. For the leaf area and nutrient evaluations, the leaves were collected on 23/11/2012 and 18/05/2013.

Plant height and diameter at breast height (DBH) data were analyzed using data from a 24-month period with measurements taken at 0, 6, 12, 18 and 24 months from planting.

3.5. Variables analyzed

3.5.1. Soil moisture

Soil moisture was determined using the gravimetric method, which consists of collecting soil samples in aluminum cans and determining the wet mass and dry mass in an oven at 105°C-110°C (for 24 hours) until constant weight and calculating the difference in weight in percentage according to Donagema *et al.* (2011). For the collection, 60 cm trenches were opened in the area of the center line. The samples were collected at three depths (0-20 cm; 20-40 cm; 40-60 cm) with three replicates of each depth collected in different profiles of the trench. The samples were placed in identified aluminum cans of known weight and then weighed in the field (wet mass determination) and subsequently taken to the 105°C-110°C oven for drying (dry mass determination) at NEPAU (Nùcleo de Ensino e Pesquisas em Agricultura Urbana), located on the Pici Campus of the Federal University of Cearà - UFC, Fortaleza, Cearà, Brazil.

3.5.2. Gas exchange

Gas exchange was analyzed using a portable infrared gas analyzer (IRGA Model LI-6400XT, Licor, USA). The readings were taken in the morning (between 8 am and 12 pm) and afternoon (between 12 pm and 3 pm). Three replications were carried out on three plants of each treatment with fully expanded leaves and good plant health under saturating light with a photosynthetically active photon flux density of approximately 1,200 μmol.m^1 .s^1 and under ambient temperature and CO_2 concentration conditions. The following variables were determined: net carbon assimilation (*A*); transpiration rate (*E*); stomatal conductance (g_s); intrinsic water use efficiency (A/g_s); momentary water use efficiency (*A/E*); leaf temperature (T_{leaf}); internal/external CO_2 concentration (C_i/C_a) and photosynthetic efficiency in the use of nitrogen (A/N) and phosphorus (A/P), expressed as μmol (CO_2) mol-^{1} (N) s-^{1} and μmol (CO_2) mol-^{1} (P) s-^{1} respectively.

3.5.3. SPAD index

The SPAD index was estimated using the portable chlorophyll meter Soil-Plant Analysis Development - SPAD-502 (Minolta Corp., Japan). Fully expanded leaves located in the middle third of the plants were used in the period between 08:00 and 10:00 in the morning. The readings were taken at five points on each leaf on either side of the vein on the adaxial side, and the value of the reading was used to calculate the average. There were five replicates per species in each water condition treatment. The SPAD index was correlated with the chlorophyll content in the leaf, but it should be noted that it is an indirect measure, as it measures the intensity of the green color which can be related to this parameter.

3.5.4. Chlorophyll-a fluorescence

A portable fluorometer (Plant Efficiency Analyser - MK2 - 9600, Hansatech, Norfolk, UK) was used to measure *chlorophyll-a* fluorescence variables. Readings were taken between 08:00 and 15:00 on fully expanded leaves at room temperature, acclimatized to the dark (with leaf clips) for 30 min. The leaves were then exposed to a saturating pulse of light for 5 s, in an area 4 mm in diameter with an arrangement of 6 LEDs at an intensity of 3.000 $\mu mol\ m^{-2}\ s^{-1}$, following the methodology suggested by Maxwell and Johnson (2000), with three replicates per species in each water condition treatment. The initial (*Fo*); maximum (*Fm*); variable (*Fv*) and maximum photosystem II efficiency (*Fv/Fm*) fluorescence variables were determined.

3.5.5. Specific leaf area and degree of succulence

To determine the leaf area, leaves were collected in two periods (dry and rainy) corresponding to November 2012 and May 2013. The plant material collected was weighed to determine fresh mass, then placed in plastic bags, packed in cooler boxes with ice and transported for subsequent determination of leaf area using the LI - 3100 equipment (Li - Cor. Inc. Lincoln, Nebraska, USA) in the Plant Physiology laboratory of the Department of Biochemistry and Molecular Biology at the Federal University of Cearà. All the material was then placed in paper bags and dried in a forced-air oven at 60° C until it reached a constant weight to determine the dry mass of the leaves.

Based on the leaf area, leaf dry mass and leaf fresh mass data, the specific leaf area (SFA, $dm^2\ g^{-1}$) and degree of succulence (GS, $g_{H2O}\ dm^{-2}$) were calculated using equations 1 and 2 respectively:

$$AFE = AF/MSF \quad (1)$$

in which,

AF - leaf area, dm2 $plant^{-1}$

DSM - leaf dry mass, g $plant^{-1}$

$$GS = (MFF - MSF)/AF \quad (2)$$

in which,

MF - fresh mass, g $plant^{-1}$

DM - fresh mass, g plant^{-1}

AF - leaf arca, dm2 plant^{-1}

3.5.6. Leaf nitrogen and phosphorus content

To determine the nitrogen and phosphorus content of the plants, two collections were made in November 2012 and May 2013, representing the dry and rainy periods respectively. The same leaves used to determine leaf area were used, and the collection procedure was described above. The plant samples of the aerial part (leaves) were dried in a forced air circulation oven ά 60°C (until constant weight) and transported to the Soil, Water and Plants Laboratory at Embrapa Agroindustria Tropical, Fortaleza, Cearà, where they were ground in a Willey mill (model MA 340, Marconi, BR), weighed on a precision scale and stored in plastic bags at room temperature and sent for analysis to the Soil, Plant Tissue and Fertilizer Chemical Analysis Laboratory in the Soils Department of the Federal University of Viçosa, Viçosa - Minas Gerais.

The solubilization of the plant tissue samples was carried out using nitric-perchloric digestion for P and sulfuric digestion for N. Nitrogen (total N) was determined by the semi-micro Kjeldahl method; phosphorus (P) by spectrometry with vanadate yellow, following the methods recommended by Embrapa (2009).

3.5.7. Growth analysis

For growth analysis purposes, plant height was measured using a 10m tape measure, from the base of the stem to the tip of the uppermost leaf, and stem diameter (after reaching a certain height, the measurement taken at 1.30m from the ground, known as Diameter at Breast Height, was adopted) using a tape measure. Four replications were carried out, with each repetition being the average of three plants, totaling 12 useful plants in the central row of the plot. The absolute growth rates in height (TCA (H), m month^{-1}) and diameter (TCA (H), cm day^{-1}) were obtained using equation 3; and the relative growth rates in height (TCR (H), m m^{-1} month^{-1}) and diameter (TCR (D), cm cm^{-1} month^{-1}) calculated using equation 4:

$$TCA = (P2 - P1)/(T2 - T1) \quad (3)$$

in which,

P = variation in plant growth

T = time variation

$$TCR = (LnP2 - LnP\ 1)/(T2 - T1) \quad (4)$$

in which,

P = variation in plant growth

T = time variation

Ln = Neperian logarithm.

3.6. Statistical analysis

The data obtained was evaluated and subjected to an analysis of variance with observations repeated over time. For the significant effects of single factors and interactions, the Tukey test was applied to compare means at 5% probability, using the SISVAR statistical software (FERREIRA, 2008).

4. RESULTS AND DISCUSSION

4.1. Soil moisture

The soil moisture profiles in the areas of each of the six species in the two water regimes (irrigated and rainfed) in the dry (November 2012) and rainy (May 2013) periods are shown in Figures 10 and 11. It was found that the soil moisture in the rainfed areas in the dry period varied between 1.0% and 2.52%, both in the marupà area (Figure 11). There was also a tendency in these areas for moisture to increase in the deeper layers, except in the guanandi area (Figure 10), which can be explained in part by the soil's physical characteristics which favor good drainage, in the surface layer, water extraction by the shallow root zone is greater than in the subsurface layers, which contributes to the maintenance of higher humidity in the deeper layers of 20 - 40 cm and 40 - 60 cm, leading to greater water conduction towards the deeper layers.

Figure 10 - Soil moisture profile in irrigated and rainfed areas planted with Mahogany (*Swietenia macrophylla* King), Guanandi (*Calophyllum brasiliense* Cambess.) and Gonçalo Alves (*Astronium fraxinifolium*) during the dry and rainy periods. Acaraù-CE, 2013.

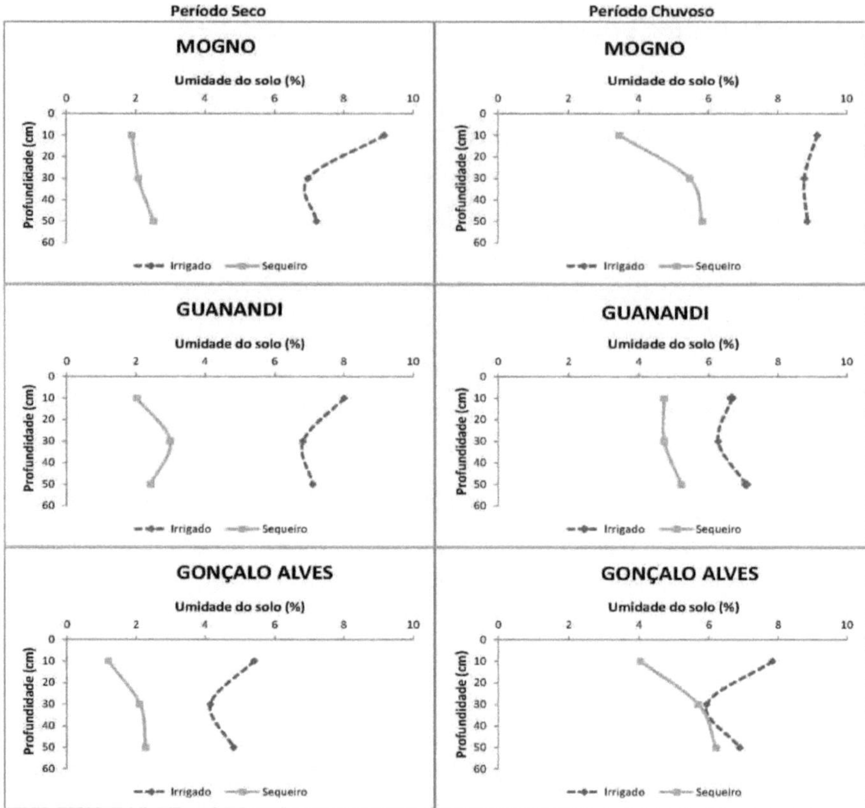

Fonte: Elaboração do autor

The fact that the guanandi area did not show greater humidity in the deepest layer (40 -60 cm) may be related to the low growth of this species in relation to the others, as it is smaller in size, which may also have reflected in less development of the deep root zone and thus had no influence on this layer under dryland conditions.

Oki (2002) in his work with *pine* found improvements in soil moisture maintenance, directly influencing the water balance of the planted area, as well as reducing the outflow of nutrients into the region's micro-basin. In relation to the irrigated area, the soil moisture profile showed the opposite trend to the rainfed area, with higher moisture levels in the top layer (0-20 cm) ranging from 5.41% in the Gonçalo Alves area (Figure 11) to 9.16% in the mahogany area (Figure 10). This higher moisture content in the surface layer is associated with the high frequency of irrigation, which favors greater moisture in the surface layer.

Figure 11 - Soil moisture profile in irrigated and rainfed areas planted with Marupâ (*Simarouba amara* Aubl.), Ipê-Amarelo (*Tabebuia serratifolia* (Vahl.) Nich.) and Ipê-Rosa (*Tabebuia impetiginosa*) during the dry and rainy periods. Acaraù-CE, 2013.

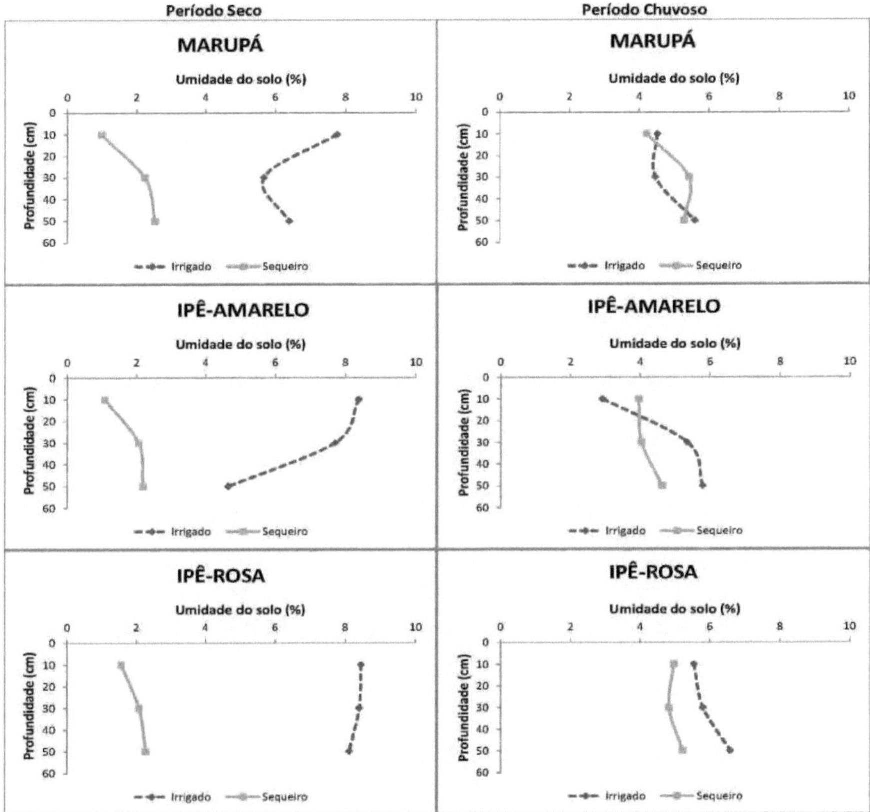

Fonte: Elaboração do autor

Vital *et al.* (1999) observed that the area planted with eucalyptus reduced runoff, providing more water into the soil, while Facco *et al.* (2012) found that eucalyptus forests regulate water runoff in micro-watersheds. Almeida and Soares (2003) found that eucalyptus plantations compare to the Atlantic forest in terms of evapotranspiration and water use in the soil, as well as efficiency in the use of water reserves in years of lower rainfall. Kennan *et al.* (2006) and Van Dijk and Kennan (2007) conclude that replacing land with conventional agriculture with forest plantations can reduce annual runoff and surface flow due to the increase in forest cover and, in some cases, can increase groundwater recharge by improving water infiltration into the soil.

Soil moisture in the dryland areas during the dry period was below the permanent wilting point of the soil, ranging from 3.11% in the 0-20 cm layer; 2.95% in the 20-50 cm layer and 3.27% in the 50-80 cm layer (Table 2), with average moisture values of around 2.16% in the mahogany area, 2.48% in the guanandi area, 1.86% in the gonçalo alves area, 1.91% in the marupà area, 1.78% in the ipê-amarelo area and 1.97 in the ipê-rosa area (Figures 10 and 11). These data suggest that these plants made greater use of water in the layers below 60 cm in order to maintain themselves, as there is a tendency for moisture to increase in the deeper layers in times of drought, as mentioned above. In the irrigated areas, soil moisture remained above field capacity, ranging from 4.02% in the 0-20 cm layer; 4.38% in the 20-50 cm layer and 4.62% in the 50-80 cm layer. The average humidity in these areas was 7.78% in the mahogany area, 7.30 in the guanandi area, 4.79% in the gonçalo alves area, 6.60% in the marupà area, 6.91% in the ipê-amarelo area and 8.31% in the ipê-rosa area (Figures 10 and 11).

In relation to the rainy season, there was an increase in the availability of water in the soil in the rainfed areas, which can be seen in the increase in humidity in all the 0-60 cm layers (Figures 10 and 11). The increase in humidity compared to the dry season was 93.9% in the mahogany area, 96.7% in the guanandi area, 186.0% in the gonçalo alves area, 160.2% in the marupà area, 147.6% in the ipê-amarelo area and 154.3% in the ipê-rosa area, which shows the great variation in soil humidity between the two seasons in the region. In the rainy season, the humidity conditions in the rainfed areas were practically the same as in the irrigated areas, except for mahogany and guanandi (Figure 10), where there was a greater difference between the areas.

The relationship between planted forests and water varies depending on the region, species, climate conditions and soil management practices employed, taking into account a micro-basin scale view (ANDRÉASSIAN, 2004; BROWN *et al.*, 2005; VAN DIJK; KEENAN, 2007). Discussions on the subject suggest that in order to adequately maintain the availability of water in the soil, management strategies are needed in the sense of previously analyzing the natural water availability conditions of the region, as well as planning irrigation. In this sense, soil moisture in regions with low-medium rainfall, as is the case in the lower Acaraù region, should be taken into account when planning the management of planted forests, given that evaporative demand is high in these areas and transpiration from these plants is an important mechanism for transferring water from the soil to the atmosphere. According to Pallardy (2008) approximately 95% of the water absorbed by the roots is released into the atmosphere by the plant as water vapor through transpiration.

The physical characteristics of the soil are also an important factor in soil water management in forest regions. Huber and Trecamam (2004) in a study with *Pinus radiata* related low transpiration rates to soils with good water retention capacity. In this study, the most obvious physical characteristic was the infiltration capacity, which comes from the texture of the soil, which varies from sand at a depth of 0 - 20 cm with an overall density of 1.51 g cm^^{1} to sandy loam at a depth of 50 - 80 cm with an overall density of 1.50 g cm^{-1} and influenced the variation in the soil moisture profile of the planted area (Table 2). Infiltration capacity is a highly important physical variable, as it directly influences groundwater recharge. According to Ilstedt *et al.* (2007) water infiltration into the soil increases on average three times more after planting trees in agricultural regions, with a restoration of the soil's hydrological processes towards a balance between evapotranspiration and infiltration rates. With high infiltration rates, the basin as a whole will act to produce water and the riparian forest will consume water, but the greater consumption will be offset by greater infiltration throughout the basin (VALENTE; GOMES, 2005).

For Bruijnzeel (2004), greater attention should be paid to the geological basis that controls the hydrological behavior of the micro-basin when analyzing the effects of changes in land use on surface flows or sediment production, and according to Van Dijk and Kennan (2007) these studies also depend on the hydrology of the region and the characteristics of the tree species. Caldato and Shumacher (2013), in a wide-ranging review of the subject, conclude that the main studies are mainly focused on *Eucalyptus grandis* in the southeast region, which is justified by the fact that it is the most important species and the region with the highest concentration of plantations, and they stress that little is known about the hydrological behavior of this species and other important species for the forestry sector in different regions of the country, and that there is much to be done in this field of research. With regard to studies of forest regions, soil moisture, in the sense of a hydrological function, is an extremely important variable for understanding physiological responses and, consequently, productivity, and can also be considered as a basis for studies on plant water deficits.

4.2. Gas exchange

4.2.1. Photosynthesis rate (A), stomatal conductance (g_s), transpiration (E) and water use efficiency

It was found that in the morning the photosynthetic rate, stomatal conductance and transpiration were influenced by the individual factors (species, water regime and season), as well as by the interaction between water regime *and* season. For the intrinsic water use efficiency variable, only the species *vs.* water regime interaction had no significant effect and for the momentary water use efficiency variable, only the single factor species and the species *vs.* water regime interaction had no significant effect (Table 3).

Table 3 - Summary of the analysis of variance for the variables: photosynthetic rate (*A*),

Stomatal conductance (g_s), transpiration (*E*), intrinsic water use efficiency (A/g_s) and momentary water use efficiency (*A/E)* in the morning in six tree species subjected to irrigated and rainfed conditions in November, February and May. Acaraù-CE, 2013.

F.V	G.L	Mean squares				
		A	g_s	*E*	A/g_s	*A/E*
Species (a)	5	113,39**	0,121**	37,35**	1533,0*	0,326 $^{n.s}$
Waste - a	12	5,96	0,007	1,68	396,2	0,203

Water regime (b)	1	565,43**	0,249**	43,08**	2132,9*	2,14**
a x b	5	27,52 $^{n.s}$	0,003 $^{n.s}$	3,66 $^{n.s}$	159,9 $^{n.s}$	0,283 $^{n.s}$
Residue - b	12	8,97	0,007	2,20	334,3	0,192
Season (c)	2	40,09*	0,130**	56,12**	15117,1**	1,314**
a x c	10	10,27 $^{n.s}$	0,010 $^{n.s}$	3,48$^{n.s}$	1160,6**	0,475**
b x c	2	127,32**	0,077**	36,88**	6378,9**	4,33**
a x b x c	10	10,21 $^{n.s}$	0,013 $^{n.s}$	2,59$^{n.s}$	964,9**	0,760**
Residue - c	48	8,17	0,007	1,84	287,9	0,116
Total	107					
CV a (%)		18,95	28,76	20,09	38,05	21,95
CV b (%)		23,24	29,13	22,97	34,95	21,31
CV c (%)		22,18	29,00	21,04	32,43	16,56

G.L - degrees of freedom; (a) - species; (b) - water regime; (c) - season; CV - coefficient of variation; **, *,$^{n.s}$ - significant by the Tukey test at 1%, 5% and not significant, respectively.

When the measurements were taken in the afternoon, there was a significant effect of the individual factors and all the interactions on photosynthetic rate and transpiration. For the stomatal conductance and momentary water use efficiency variables, only the triple interaction between the factors had no significant effect. Intrinsic water use efficiency was only influenced by the factors water regime and season alone (Table 4).

Table 4 - Summary of the analysis of variance for the following variables: photosynthetic rate (*A*), stomatal conductance (*gs*), transpiration (*E*), intrinsic water use efficiency (*A/gs*) and momentary water use efficiency (*A/E)* in the afternoon in six tree species subjected to irrigated and rainfed conditions in November, February and May. Acaraù-CE, 2013.

F.V	G.L	Mean squares				
		A	*gs*	*E*	*A/gs*	*A/E*
Species (a)	5	181,48**	0,150**	66,59**	783.19 $_{n.s}$	0,822**
Waste - a	12	4,74	0,004	0,89	362,84	0,145
Water regime (b)	1	1064,38**	0,207**	70,23**	1372,08*	7,44**
a x b	5	38,58**	0,026*	21,32**	451,18 $^{n.s}$	0,469*
Residue - b	12	6,31	0,006	1,41	242,48	0,120
Season (c)	2	117,65**	0,264**	101,60**	4731,28**	1,933**
a x c	10	17,92**	0,014**	3,57*	549,76 $^{n.s}$	0,481*
b x c	2	291,73**	0,156**	28,91**	116,70 $^{n.s}$	2,216**
a x b x c	10	12,01*	0,007 $^{n.s}$	4,07**	239,15 $^{n.s}$	0,237 $^{n.s}$
Residue - c	48	5,83	0,004	1,42	370,90	0,183

Total	107					
CV a (%)		17,00	23,54	12,66	40,14	22,58
CV b (%)		19,61	26,92	15,92	32,81	20,54
CV c (%)		18,92	23,59	15,93	40,58	25,35

G.L - degrees of freedom; (a) - species; (b) - water regime; (c) - season; CV - coefficient of variation; **, *,[n.s] - significant by the Tukey test at 1%, 5% and not significant, respectively.

Figure 12 shows the variation in the rate of photosynthesis in irrigated and rainfed plants as a function of the evaluation periods in the morning (A) and afternoon (B). Taking into account the interaction between water regime *and* season in the morning period, it can be seen that in November and February CO_2 assimilation rates were significantly affected in rainfed plants, with reductions of 75.74% in November and 75.57% in February, due to the water stress caused by the dry season. Later in the rainy season, the rainfed plants recovered in response to the availability of water in the soil. In the afternoon, the response pattern is similar, but with greater perceptions of a drop in the rate of photosynthesis in the dry season, with reductions of 110.95% and 122.75% in November and February respectively.

Figura 12 - Photosynthetic rate (*A)* in six tree species subjected to irrigated and rainfed conditions as a function of time in the morning (A) and afternoon (B). Acaraù-CE, 2013.

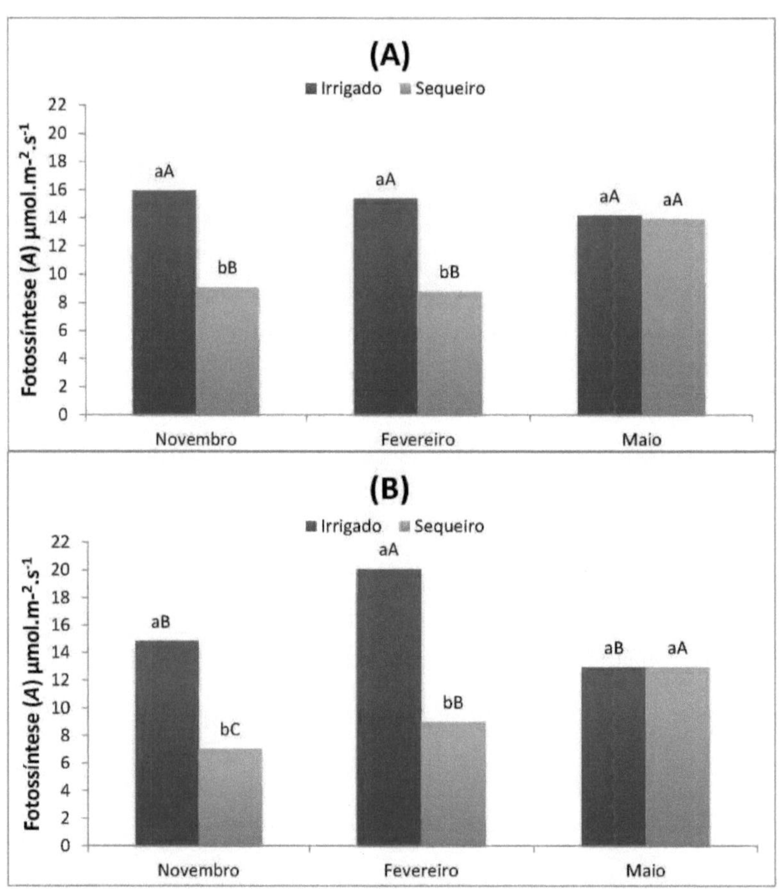

Averages followed by the same letter do not differ from each other at 5% probability, using the Tukey test. Lowercase letters (between water regime treatments) and uppercase letters (between seasons within the same water regime) Source: Author's elaboration.

In relation to the effect of the seasons on CO_2 assimilation rates, it can be seen that the irrigated plants did not significantly alter *A* in the different seasons, which can be explained by the availability of water in the soil, which did not vary in a contrasting way for these plants. However, in the month of February, in the afternoon, high *A* values were observed, which can be attributed to the solar radiation conditions resulting from this season. In rainfed plants, the distribution of rainfall during the year has an influence on photosynthesis rates under these conditions, so that they show lower values during the dry season and a significant increase during the rainy season. Photosynthesis reaches its lowest values in November, with 7.03 µmol m^{-2} s^{-1} observed in the afternoon.

Figure 13 shows the variation in the rate of photosynthesis taking into account the three-way interaction between the factors, representing the interaction between species *and* season in the two water regimes in the afternoon. It can be seen that, with the exception of guanandi, all the species under irrigated conditions obtained higher *A* values in February. As the availability of water in the soil was not a limiting factor, another intrinsic factor ά at this time may have influenced this higher peak in CO_2 assimilation, probably light intensity or photosynthetically active radiation. In rainfed plants, there was greater variation in photosynthesis rates, both in relation to the seasons and in the behavior of the different species. Mahogany and guanandi showed lower *A* values in November and February, corresponding to the dry season. The species ipê-marelo and ipê-rosa had significantly lower rates only in November, during the driest period, while the species gonçalo alves and marupà did not significantly alter their photosynthesis rates. With regard to the variation in *A,* the six different species showed three different response patterns according to the different rainfall periods.

The sharpest reduction in photosynthetic capacity was seen in guanandi and mahogany, with values of 0.46 and 1.82 μmol m^{-2} s^{-1} , respectively, observed in February. The degree of intensity of the water stress was more pronounced in these species as they are typical of the humid regions of the Amazon, demonstrating their susceptibility to the environment of little water in the soil provided by the rainfed environment. In the case of guanandi, partial dehydration of the leaves was observed, with some plants becoming yellowish and with a shriveled appearance. Despite having been affected by water stress at different times, the ipê-amarelo and ipê-rosa showed a recovery in photosynthetic capacity in February, towards the end of the drought. This recovery may be associated with the small amount of rain observed at the beginning of 2013, which, despite the small amount, may have been enough for these species to develop mechanisms to increase CO_2 assimilation rates, such as an increase in leaf area, which is characteristic of Caatinga plants. Unlike the other species, gonçalo alves and marupà did not show a significant reduction in *A* rates as soil water availability varied, which suggests that these species have more efficient physiological mechanisms for tolerating dry conditions in relation to their photosynthetic capacity.

Figura 13 - Photosynthetic rate (A) in six tree species subjected to irrigated and rainfed conditions in November, February and May in the afternoon. Acaraù-CE, 2013.

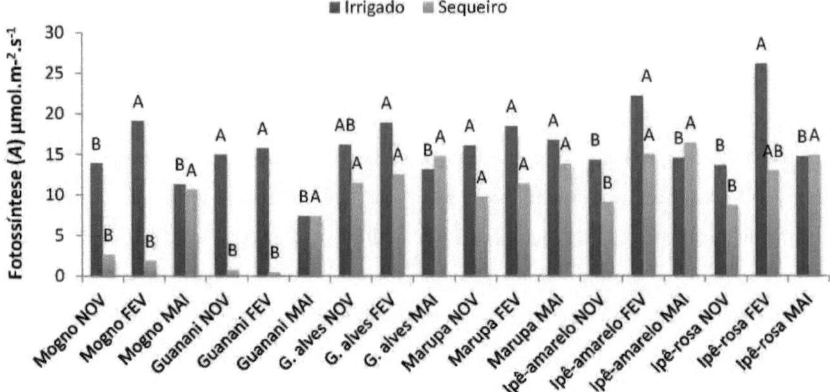

Averages followed by the same letter do not differ from each other at 5% probability, using the Tukey test. Source: Prepared by the author.

The response of these species is similar to that obtained in plants of *Khaya ivorensis* (ALBUQUERQUE *et al.*, 2013), *Carapa guianensis* (GONÇALVES *et al.*, 2009), *Minquartia guianensis* (LIBERTAO *et al.*, 2006) and in *eucalyptus and acacia spp.* (WARREN *et al.*, 2011) in *Swietenia macrophylla* and *Tabebuia serratifolia* (CORDEIRO, 2012) and

Anacardium occidentale L. at certain times of the year (LIMA *et al.*, 2010). The reduction in *A*, in plants subjected to water deficit, occurs for various reasons, such as stomatal limitation to CO_2 flow, damage to the photochemical apparatus of photosynthesis, reduction in ATP synthesis, and reduction in the activity of Rubisco (Ribulose-1,5-biphosphate carboxylase oxygenase) or in its regeneration rate (FLEXAS; MEDRANO, 2002).

According to Larcher (2006), photosynthetic capacity changes seasonally as a result of adaptations to environmental conditions. Perennial plants from regions with alternating periods of rain and drought produce morphologically distinct leaves and photosynthetic behavior depending on the season. Mendes *et al.* (2013) work with *Cordia oncocalyx* well characterizes this change in physiological responses during a few days after the end of the rainy season in semi-arid Brazil.

The results show that the response patterns of the carbon assimilation rate are different for the species, with a limitation of the photosynthetic work caused by environmental factors in the different seasons, with the most negative effects being observed in species native to the Amazon (mahogany and guanandi). The other species showed better adaptation to conditions of low water availability in the soil during the dry season. This adaptation is not adjusted to the highest capacity, but rather to the best relationship between gains and risks. This response pattern is described by Blackmam (1905). Palhares *et al.* (2010), working on the photosynthetic responses of serrado plants in the dry and rainy seasons, found that in the dry period, there are species that depress the photosynthetic rate of carbon by up to 50%, while in other species it remains unchanged, and reinforces this by relating it to adaptive characteristics such as the reduction in leaf area and the storage of reserves in the stem and underground system. With regard to photosynthetic responses between species, the genetic differences are quite considerable. The causes of different photosynthetic capacities lie in the anatomical peculiarities of the leaves (the ease with which air can diffuse into the intercellular spaces, the shape and distribution of the stomatal apparatus) and in the efficiency and quantity of carboxylation enzymes (LARCHER, 2006).

In relation to stomatal conductance as a function of water regime in the different seasons, Figure 14 shows that in the months of November and February, referring to the dry season, plants in rainfed conditions showed lower stomatal conductance values compared to plants in irrigated conditions, with a reduction of 110.95% and 56.84% in the morning and 82.23% and 86.49% in the afternoon in the respective months of November and February. These results show that these plants promote stomatal closure to reduce water loss when exposed to water deficit conditions.

Under stress conditions, especially water and salt stress, stomatal closure can be seen as a positive response by the plant to maintain water (TAIZ; ZEIGER, 2013). These results are similar to those found by Tonello and Filho (2012) who found a decrease in stomatal conductance values in *Pterogyne nitens*, *Aspidosperma polyneuron* and *Myroxylum peruiferum* when subjected to different water availability in the soil. According to Chaves *et al.* (2004), plants that reduce the opening of stomata in situations of water deficit are more conservative in their use of water. These same authors evaluated eucalyptus clones in irrigated and rainfed conditions and observed no significant difference between the water treatments in relation to stomatal conductance. In the rainy season, no significant difference was observed between the regimes, demonstrating that there was no limitation in the availability of water in the soil during this period and that this did not influence g_s.

Figura 14 - Stomatal conductance (g_s) in six tree species subjected to irrigated and rainfed conditions as a function of time in the morning (A) and afternoon (B). Acaraù- CE, 2013.

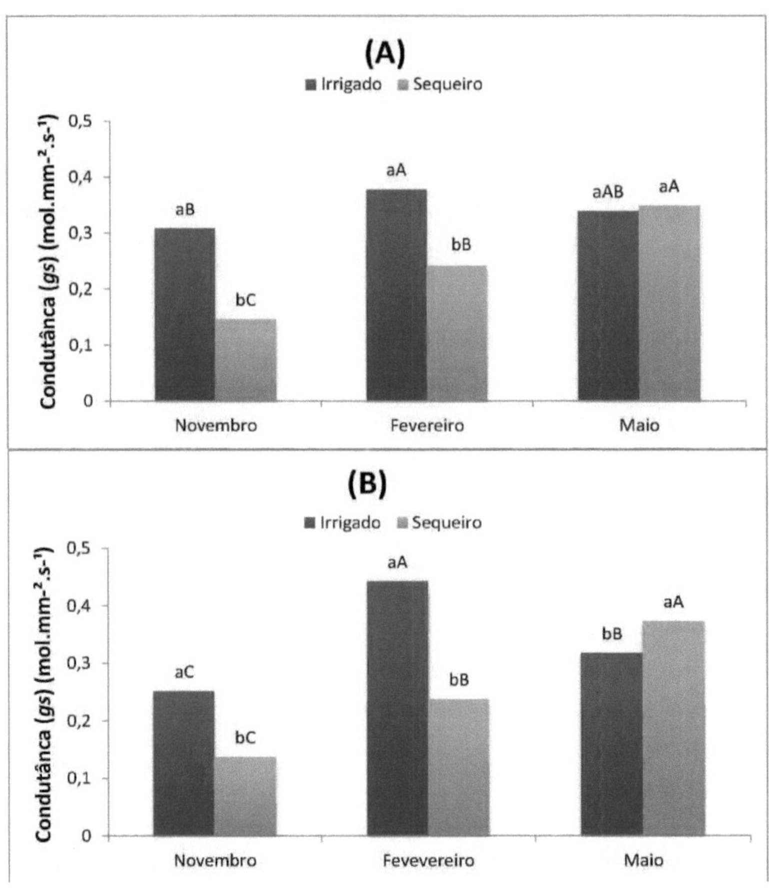

Averages followed by the same letter do not differ from each other at 5% probability, using the Tukey test. Lowercase letters (between water regime treatments) and uppercase letters (between seasons within the same water regime) Source: Author's elaboration.

With regard to the variation in stomatal conductance over the seasons (Figure 14) in the two different regimes, there was an increase in *gs* in the plants irrigated in February in both shifts. This variation may have been influenced by environmental factors but also by irrigation management which is suspended in the rainy season, which may explain the drop in *gs* values in the irrigated conditions. In rainfed plants, *gs* was significantly influenced, increasing according to the water availability conditions inherent to the seasons. The reduction in *gs* was 170.48% when comparing the wettest season to the driest season, showing the stomatal regulation occurring in the plants in response to a better water condition in the soil. For irrigated plants, this reduction was 76.20% in the most extreme case.

It can clearly be seen that the effect on dryland plants was much more intense given the severity of the stress. The small reduction in irrigated plants between the periods may have been caused by conditions characteristic of the dry season, such as the greater vapor pressure deficit influencing *gs* even in conditions of good water availability in the soil. Silva *et al.* (2008) working with *Schinus terebinthifolius* observed a reduction in *gs* in this plant when subjected to different water regimes. Rodrigues *et al.* (2011) when investigating the effect of climatic seasonality on stomatal conductance in *Jatropha curcas* in semi-arid conditions, found a decrease in *gs* at times of lower rainfall, and observed that in addition to the availability of water in the soil, other factors such as air temperature and photosynthetically active radiation influenced the physiological responses of this species. In regions where the seasonality of rainfall does not alter soil humidity, *gs* suffers little change, as reported by Mendes and Marenco (2010) in a study of ten native species in the Central Amazon region, although maximum photosynthesis rates are minimal in the dry season. According to Larcher (2006), the degree of stomatal aperture continually adjusts to fluctuations in environmental factors. In vascular plants, water deficiency primarily affects the stomatal apparatus, even though dry air causes the stomata to close, water deficiency in the soil can act at the same time, amplifying this effect.

With regard to transpiration, Figure 15 shows the unfolding of the interaction between water regime *and* season in the morning (A) and afternoon (B) shifts. Similarly, in both shifts, transpiration rates were significantly reduced under rainfed conditions in the dry seasons (November and February), with more significant reductions in the afternoon shift, with values 65.67% and 38.56% lower in November and February, respectively, compared to irrigated conditions.

Figura 15 - Transpiration rate *(E)* in six tree species subjected to irrigated and rainfed conditions as a function of time in the morning (A) and afternoon (B). Acaraù- CE, 2013.

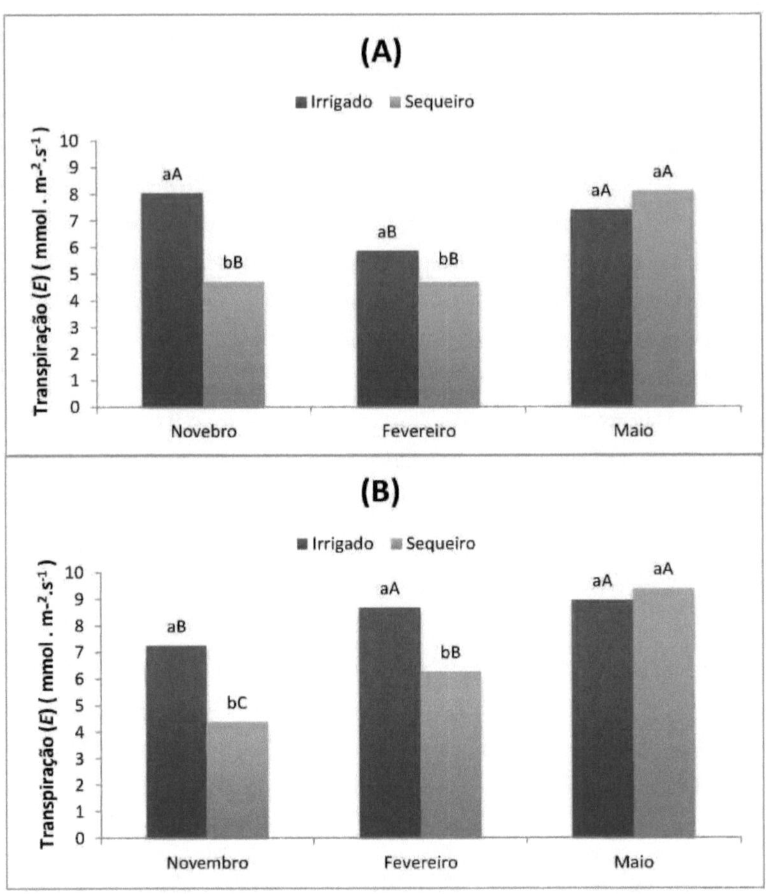

Averages followed by the same letter do not differ from each other at 5% probability, using the Tukey test. Lowercase letters (between water regime treatments) and uppercase letters (between seasons within the same water regime) Source: Author's elaboration.

In May, there was an increase in *E* in response to the greater water availability at that time, with no significant difference between the two water regimes. Variations in transpiration rates followed the same trend as stomatal conductance Figure 14 in plants subjected to the rainfed regime, which shows that plants reduce water loss by transpiration through stomatal closure according to the conditions of water availability. Irrigated plants showed a drop in transpiration despite high conductance, which may have been influenced by lower temperatures in the morning, but in the afternoon *E* increased. These results corroborate Cordeiro *et al.* (2009) who found lower transpiration rates in *Swietenia macrophylla* plants when subjected to water deficiency, concluding that this species is efficient at controlling transpiration rates. This behavior is also similar to that found by Scalon *et al.* (2011) when working with *Guazuma ulmifolia* , where transpiration rates were lower when subjected to a lower water supply. Nogueira and Silva (2002) found this same effect in *Schinopsis brasiliensis* after suspending irrigation.

Figure 16 shows that taking into account the three-way interaction between the factors considering the species *vs.* season interaction in the two water regimes in the afternoon shift, it can be seen that under irrigated conditions, mahogany and gonçalo alves did not significantly alter *E* in the seasons. Marupà, ipê-amarelo and ipê-rosa showed higher transpiration rates in the rainy season in May, when there is more water available in the soil, while guanandi showed the highest *E* rate at the end of the dry season in February. With regard to the plants under rainfed conditions, it was observed that mahogany, guanandi and ipê-amarelo showed significantly lower rates of *E* in the drier months (November and February) than in May, suggesting that these species are more efficient at saving water by controlling stomatal closure and reducing transpiration rates. Neto (2003) found this reduction throughout the year in *Mangifera indica*, where the lowest values were observed during periods of water stress.

Reducing transpiration in plants subjected to water deficit helps to conserve available water. This reduction can occur through the temporary closure of stomata, as a modulative adaptation. A modifying change occurs when leaves growing under conditions of water deficiency have smaller stomata but a higher stomatal density. This modification provides conditions for a faster reduction in transpiration by regulating stomatal closure (LARCHER, 2006). By regulating stomatal aperture, the plant is able to modulate transpiration rates according to the possibilities and needs of its water balance, a fact not investigated in this study. The control of gas exchange through stomatal closure is considered a complex process, as plants face a dilemma between the decrease in CO_2 assimilation through stomatal closure and the loss of water through opening, both of which are negative for plant development, although the tendency is to favor photosynthetic assimilation (PIMENTEL, 1998; ANGELOCCI, 2002).

Figura 16 - Transpiration rate *(E)* in six tree species subjected to irrigated and rainfed conditions in November, February and May in the afternoon. Acaraù- CE, 2013.

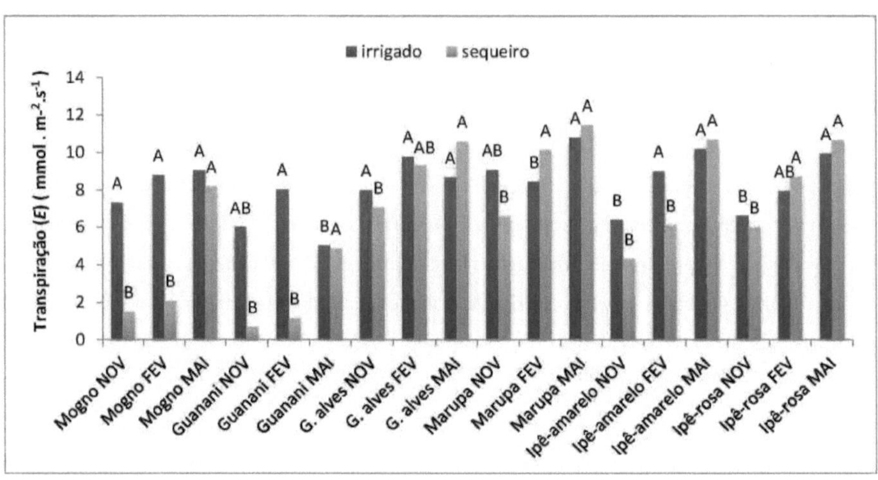

Averages followed by the same letter do not differ from each other at 5% probability, using the Tukey test. Source: Prepared by the author.

The gonçalo alves, marupà and ipê-rosa showed lower rates of *E* in November, but an increase at the end of the drought in February, so they did not differ significantly from the rainy season, suggesting that stomatal closure is less effective and less efficient in conserving water at the end of the drought in these plants compared to the other species. Some authors have observed this behavior in other tree species, Delgado (2012) observed that the species *Peltophorum dubium* did not alter water loss by transpiration when subjected to different water management and Tatagiba *et al.* (2008) working with eucalyptus clones in environments with different soil water availability, found that there was no significant difference in transpiration.

As for water use efficiency, Figure 17 shows the variation in intrinsic efficiency (A/gs) in the different seasons during the morning shift, considering the three-way interaction between the factors. It can be seen that in plants under irrigated conditions, A/gs did not change significantly depending on the season, as the amount of water was not limiting in these conditions. In plants under dry conditions, gonçalo alves and ipê- rosa did not change their A/gs depending on the season, while the other species had higher values in the driest month (November).

Figura 17 - Intrinsic water use efficiency (A/gs) in six tree species subjected to irrigated and rainfed conditions in the months of November, February and May in the morning. Acaraù-CE, 2013.

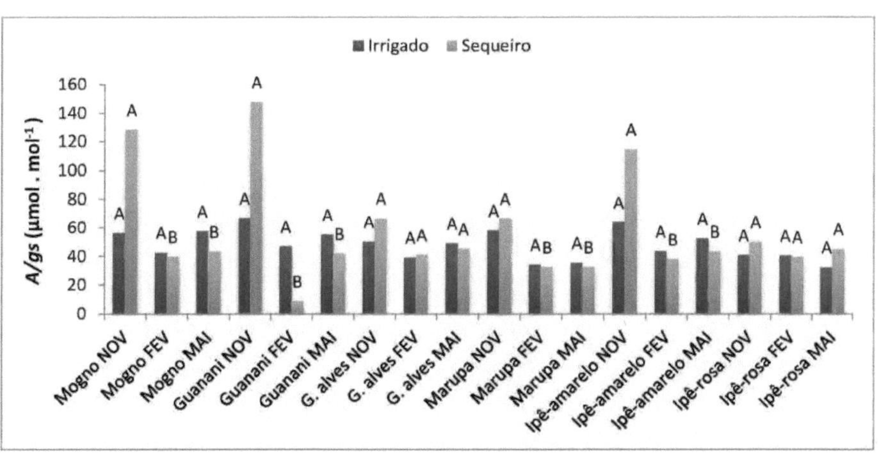

Averages followed by the same letter do not differ from each other at 5% probability, using the Tukey test. Source: Prepared by the author.

The greater efficiency of water use during the dry season shows that these species improve water management within the plant during periods of lower water availability due to the greater stomatal closure observed during this period (Figure 14), while on the other hand decreasing CO_2 assimilation during the same period (Figure 12). According to Taiz and Zeiger (2013), water stress tends to reduce *gs* first before reducing the photosynthetic rate, so it is possible that the plant assimilates more CO_2 molecules for each unit of water transpired, making it more efficient at using the available water.

Analyzing the intrinsic efficiency of water use (*A/gs*) in the afternoon shift, Figure 18 shows the effect of the season factor in isolation. It can be seen that the plants increase *A/gs* according to the water availability of the season. Water use, taking *A/gs* into account, was on average 48.45% higher in the driest month than in the wettest month.

Figura 18 - Intrinsic water use efficiency (*A/gs)* in six tree species subjected to irrigated and rainfed conditions in November, February and May in the afternoon. Acaraù-CE, 2013

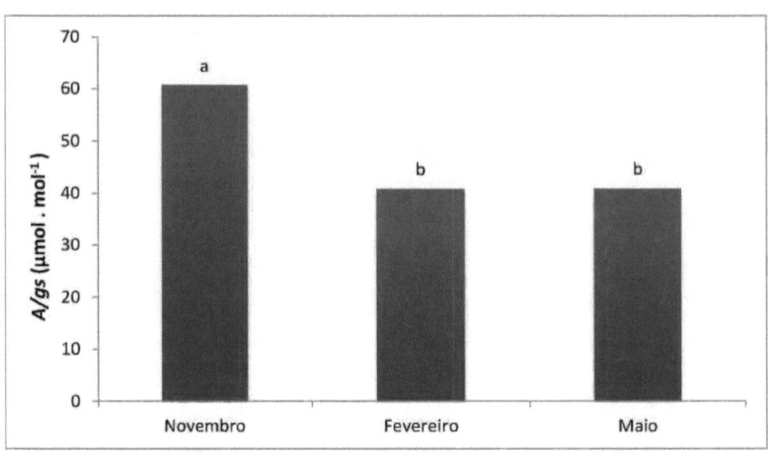

Averages followed by the same letter do not differ from each other at 5% probability, using the Tukey test. Source: Prepared by the author.

Figura 19 Momentary water use efficiency *(A/E)* in six tree species subjected to irrigated and rainfed conditions in November, February and May in the morning. Acaraù-CE, 2013

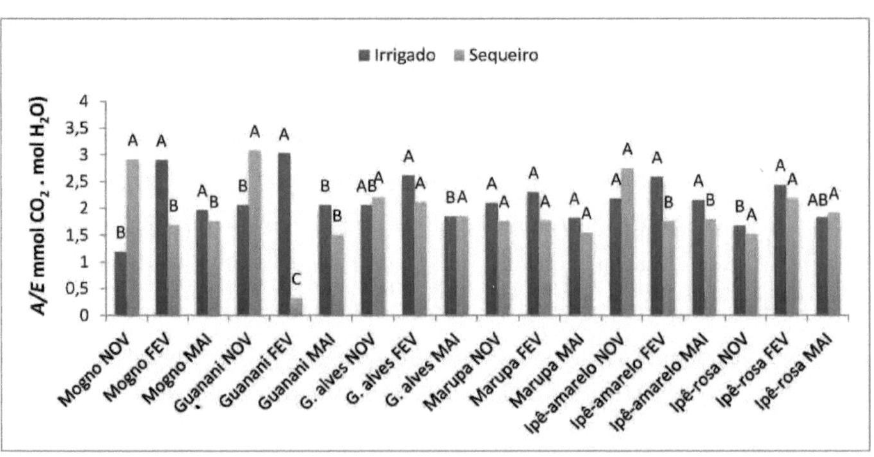

Averages followed by the same letter do not differ from each other at 5% probability, using the Tukey test. Source: Prepared by the author.

In relation to the momentary efficiency of water use *(A/E)* in the morning shift (Figure 19), it was found that the plants under irrigated conditions had a higher *A/E* in February, as a result of the low transpiration rates observed in that month in the morning (Figure 15), which helped with internal water conservation. In relation to the rainfed plants, mahogany, guanandi and ipê-amarelo had significantly higher *A/E* averages in the driest month (November), while gonçalo alves, marupà and ipê- rosa did not significantly alter *A/E*, in which case water restriction did not significantly change CO_2 assimilation per unit of transpired water.

Figure 20 shows the momentary efficiency of water use *(A/E),* taking into account the interaction between water regime *and* season. Analyzing the driest month when the water in the soil is more contrasting between the two periods, it was found that in the morning shift the plants under rainfed conditions have a higher *A/E* than the irrigated plants, but in the afternoon it drops to a lower level. In this case, the drop in CO_2 assimilation contributes more effectively than the reduction in transpiration, reducing the *A/E* ratio. In the rainy season, when the water conditions are favorable to both regimes, the ratio equals out. As for the variation in the *A/E* of the regimes in the seasons, it can be seen that the daily course modifies the *A/E* in rainfed plants, so that in the morning it is higher in the drier season, but in the afternoon shift there were no significant changes in *the A/E* ratio. The variation in the rate of CO_2 assimilation (*A*) and transpiration rate (*E*) during the day has an influence depending on the condition of water availability, so that in the dry season *A* is more effective in this fluctuation, while in the rainy season *E is* more influential in this process of *A/E* variation.

In relation to the plants subjected to irrigated conditions in which there was no water restriction in the soil, the values of (*A/gs*) and *(A/E)* were lower than those obtained in other tropical tree species (NINA JUNIOR, 2009; ATROCH, 2008; MARRENCO *et al.*, 2001). According to Larcher (2006), this physiological parameter varies between and within plant species, a fact that was verified by Pinzón-Torres and Schiavinato (2008) in four tropical leguminous tree species. With regard to rainfed plants, the fact that *(A/E)* did not increase with water deficiency in gonçalo alves, marupà and ipê-rosa indicates that the reduction in stomatal conductance did not work efficiently to reduce water loss via stomata at first in these species. However, taking the seasons into account, these plants in the driest season showed higher rates of CO_2 assimilation per unit of water transpired in the morning, showing water-conserving behavior in times of low water availability in the soil (Figure 20-A).

Figura 20 - Momentary water use efficiency *(A/E)* in six tree species subjected to irrigated and rainfed conditions as a function of time in the morning (A) and afternoon (B). Acaraù-CE, 2013.

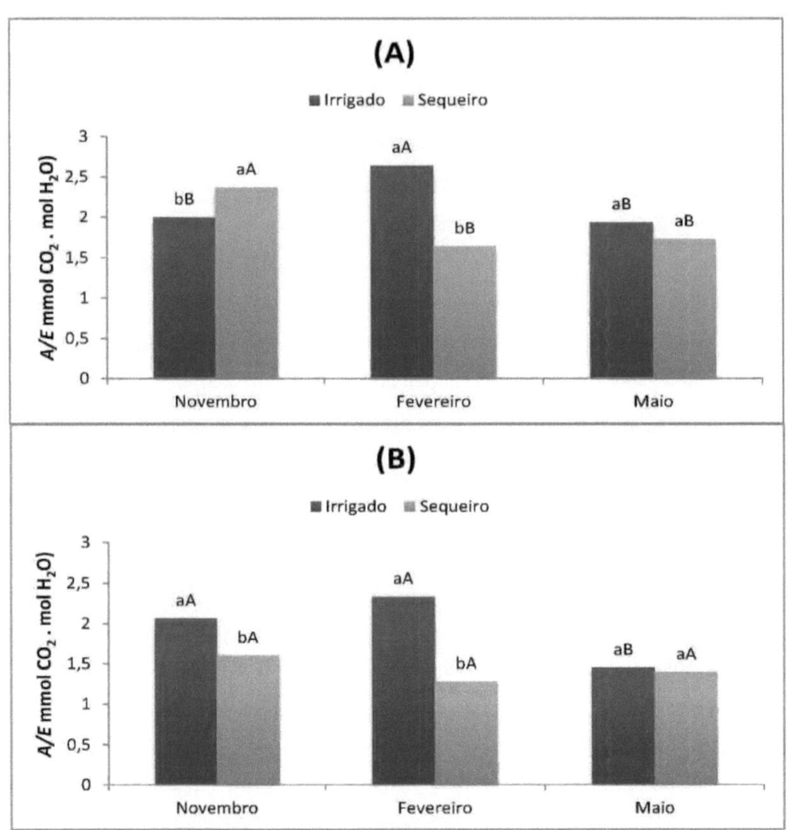

Averages followed by the same letter do not differ from each other at 5% probability, using the Tukey test. Lowercase letters (between water regime treatments) and uppercase letters (between seasons within the same water regime) Source: Author's elaboration.

Water use efficiency always changes when the conditions for CO_2 or H_2O diffusion are altered. When stomata are fully open, CO_2 absorption is more limited by transfer resistance than water loss through transpiration. The best ratio between CO_2 absorption and H_2O loss is achieved when the stomata are partially closed. This situation can be demonstrated at the beginning of water deficiency, when both diffusion processes are rapidly reduced and *(A/E)* reaches its highest values. When the stomata are practically closed, the *(A/E)* ratio declines rapidly, as the entry of CO_2 is more restricted than stomatal transpiration and water loss in the form of vapor continues to occur through the cuticle (LARCHER, 2006).

Evaluating the response of a *Pinus pinaster* plantation to drought, Jarosz *et al.* (2008) concluded that water use efficiency increases in the driest season with increasing soil water deficit and decreases with increasing vapor pressure deficit. This response was also observed in *Jatropha curcas* when subjected to lower levels of soil water deficit (ROZA, 2010), in *Ricinus communis* under severe water stress (Ψw = -1.6 to -2.1 MPa) (SAUSEN, 2007) and in *Stryphnodendron adstringens* where maximum *A/E* values were observed when leaf water potential values were greatly reduced 27 days after the start of the water deficit (ROCHA; MORAES, 1997).

4.2.2 Relationship between internal and external CO_2 concentration (Ci/Ca) and leaf temperature (Tleaf)

Table 5 shows the analysis of variance for the variables internal/external CO_2 ratio (*Ci/Ca*) and leaf temperature (Tleaf) during the morning shift. It can be seen that only the species *vs.* water regime interaction did not significantly influence the *Ci/Ca* ratio. For the leaf temperature variable, the single factor species and the interaction species *vs.* water regime had no significant effect. In the afternoon shift, Table 6 shows that all the factors and their interactions had a significant effect on the *Ci/Ca* ratio. For leaf temperature, the single factors species and season, as well as the interactions species *vs.* season and water regime *vs.* season were significant.

Figure 21 shows the unfolding of the three-way interaction between the factors in the morning (A) and afternoon (B) shifts for the *Ci/Ca* ratio. It was found that in the morning the plants under irrigated conditions did not change the *Ci/Ca* as a function of the seasons. In the plants subjected to rainfed conditions, only gonçalo alves and ipê-rosa showed no significant change in *Ci/Ca*, while the other species in these conditions showed lower values in the drier season. This may be related to the lower stomatal aperture in the dry season, in response to water deficiency, reducing CO_2 absorption.

Table 5 - Summary of the analysis of variance for the variables: internal/external CO_2 ratio (*Ci/Ca*) and leaf temperature (Tleaf) in the morning in six tree species subjected to irrigated and rainfed conditions in November, February and May. Acaraù-CE, 20R

F.V	G.L -	Q.M	
		Ci/Ca	Tleaf
Species (a)	5	0,022*	0.404 n.s
Waste - a	12	0,006	0,306
Water regime (b)	1	0,034*	29,984**
a x b	5	0,003 n.s	0,645 n.s
Residue - b	12	0,004	0,294
Season (c)	2	0,298**	169,107**
a x c	10	0,015**	1,426**
b x c	2	0,104**	2,499**
a x b x c	10	0,019**	0,955*

Residue - c	48	0,005	0,434
Total	107		
CV a (%)		10,96	1,64
CV b (%)		9,60	1,61
CV c (%)		10,52	1,96

G.L - degrees of freedom; (a) - species; (b) - water regime; (c) - season; CV - coefficient of variation; **, *,[n.s] - significant by the Tukey test at 1%, 5% and not significant, respectively.

Table 6 - Summary of the analysis of variance for the variables: internal/external CO_2 ratio (*Ci/Ca*) and leaf temperature (T_{leaf}) in the afternoon in six tree species subjected to irrigated and rainfed conditions in November, February and May. Acaraù-CE, 2013.

F.V	G.L	Q.M	
		Ci/Ca	Tleaf
Species (a)	5	0,0180**	2,231**
Waste - a	12	0,0011	0,189
Water regime (b)	1	0,0997**	0,777 [n.s]
a x b	5	0,0320**	0,487 [n.s]
Residue - b	12	0,0007	0,255
Season (c)	2	0,0407**	7,160**
a x c	10	0,0173**	0,763*
b x c	2	0,0049*	2,721**
a x b x c	10	0,0115**	0,596 [n.s]
Residue - c	48	0,0014	0,327
Total	107		
CV a (%)		4,55	1,23
CV b (%)		3,75	1,43
CV c (%)		5,23	1,62

G.L - degrees of freedom; (a) - species; (b) - water regime; (c) - season; CV - coefficient of variation; **, *,[n.s] - significant by the Tukey test at 1%, 5% and not significant, respectively.

These results corroborate Albuquerque *et al.* (2013) who observed a 37% reduction in the *Ci/Ca* ratio in *Khaya ivorensis* plants when subjected to water stress. Cordeiro (2012) also observed a drop in the *Ci/Ca* ratio in *Swietenia macrophylla* and *Tabebuia serratifolia* plants during the period of lowest rainfall. Tomczac (2012) working with *Tabebuia impetiginosa* found a reduction in internal CO_2 concentration when subjected to water stress in two different soil types.

Figura 21 Internal/external CO_2 ratio (*Ci/Ca*) in six tree species subjected to irrigated and rainfed conditions in November, February and May in the morning (A) and afternoon (B). Acaraù-CE, 2013

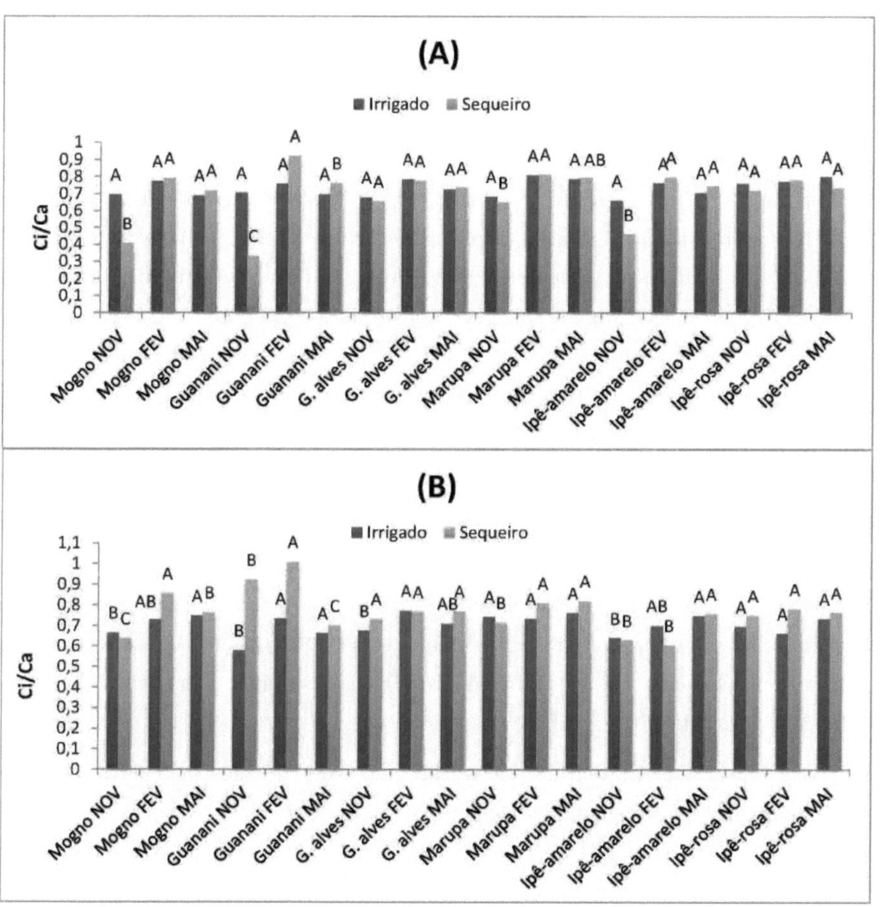

Averages followed by the same letter do not differ from each other at 5% probability, using the Tukey test. Source: Prepared by the author.

According to Larcher (2006), when plants are subjected to water stress, they reduce CO_2 intake due to stomatal closure and, as a consequence, a reduction in internal concentration, demonstrating an interdependence between CO_2 assimilation and water consumption. In the afternoon shift (Figure 21-B) it can be seen that the *Ci/Ca* ratio in the two water regime conditions is influenced by falls in the rate of CO_2 assimilation (*A*) and stomatal conductance (*gs*), especially in the drier season (November). With the exception of marupà and ipê-rosa, all the species in irrigated conditions showed lower *Ci/Ca* values in the driest period. Under rainfed conditions, gonçalo alves and ipê-rosa maintained *Ci/Ca* at significantly the same levels in relation to the seasons, suffering less influence on the variation of *A* and *gs* between shifts than the other species. Mahogany and guanandi had the highest *Ci/Ca* in February, which can be explained by the significant increase in stomatal conductance followed by a sharp drop in the rate of CO_2 assimilation in these species (Figure 13). Marupà and ipê-amarelo had higher mean *Ci/Ca* in the wetter season and lower in the drier season, following the trend of increased CO2 assimilation.

According to Larcher (2006), CO_2 pressure in the intercellular spaces is defined by the balance between consumption (photosynthesis) and replacement (inflow from the external environment, respiration, photorespiration) and stomata often react to this balance, This is because stomatal resistance changes as a function of photosynthetic intensity, maintaining proportionality between the two processes, and as long as this optimizing trend is maintained, the *Ci/Ca* ratio remains constant. However, the lower rate of photosynthesis (*A*) observed in the dry period can be attributed to both stomatal limitations, through the availability of CO_2 in the mesophyll and carboxylation sites; and non-stomatal limitations, possibly determined by the partial inactivation of rubisco as well as the quantity of this enzyme (RIBEIRO; MACHADO, 2007).

Figure 22 shows the variation in leaf temperature in the different seasons during the morning shift. It can be seen that at *Tleaf* in the morning, both irrigated and rainfed plants showed significantly higher leaf temperatures in the driest season (November). In irrigated plants, leaf temperatures were lower in February (the end of the dry season) than in the rainy season (May), in which case the irrigation applied to this treatment caused the plants to lower their *Tleaf* even during a dry period, a fact not observed in rainfed conditions.

Figura 22 - Leaf temperature (*Tleaf*) in six tree species subjected to irrigated and rainfed conditions in November, February and May in the morning. Acaraù-CE, 2013

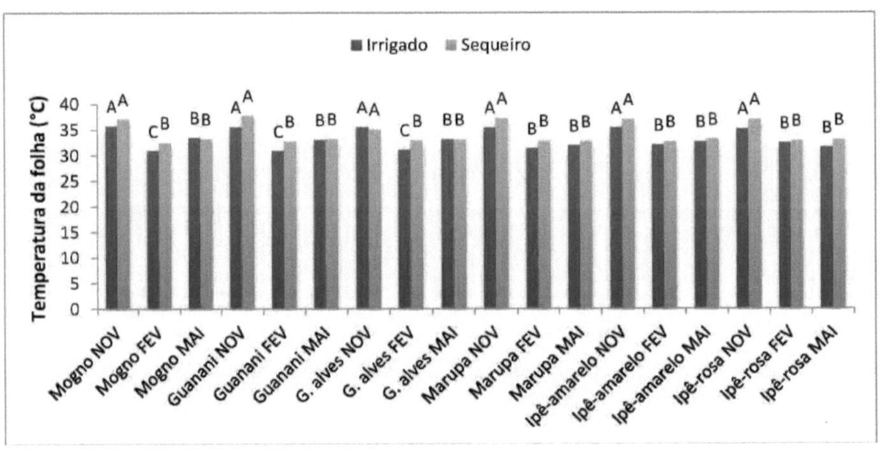

Averages followed by the same letter do not differ at 5% probability, using the Tukey test. Source: Prepared by the author.

Analyzing the two-way interaction between water regime *and* season in both shifts (Figure 23), it can be seen that during the dry season, irrigated plants maintained a lower leaf temperature than rainfed plants in both shifts. Such plants under rainfed conditions only show leaf temperature regulation during the rainy season.

These results show that the more hydrated plants promoted better temperature control in the leaf tissue, which may also be related to the higher transpiration rates. In addition to factors such as the degree of hydration promoted by the availability of water in the soil and the rate of transpiration, environmental factors such as air temperature, vapor pressure deficit (VPD) and solar radiation may have influenced the differences found in the seasonal factor, since the environmental characteristics are quite different between the seasons.

Figure 24 illustrates the difference between leaf and air temperature in the three seasons during the morning and afternoon shifts. It can be seen that the irrigated plants tended to maintain a lower leaf temperature than the air temperature regardless of the time of day. However, the rainfed plants showed a different dynamic, with leaf temperatures higher than the air temperature during the dry season, with this temperature only regulating with the availability of water in the soil at the end of the dry season and the beginning of the rainy season.

Figura 23 - Leaf temperature (*Tieaf*) in six tree species subjected to irrigated and rainfed conditions as a function of time in the morning (A) and afternoon (B). Acaraù- CE, 2013.

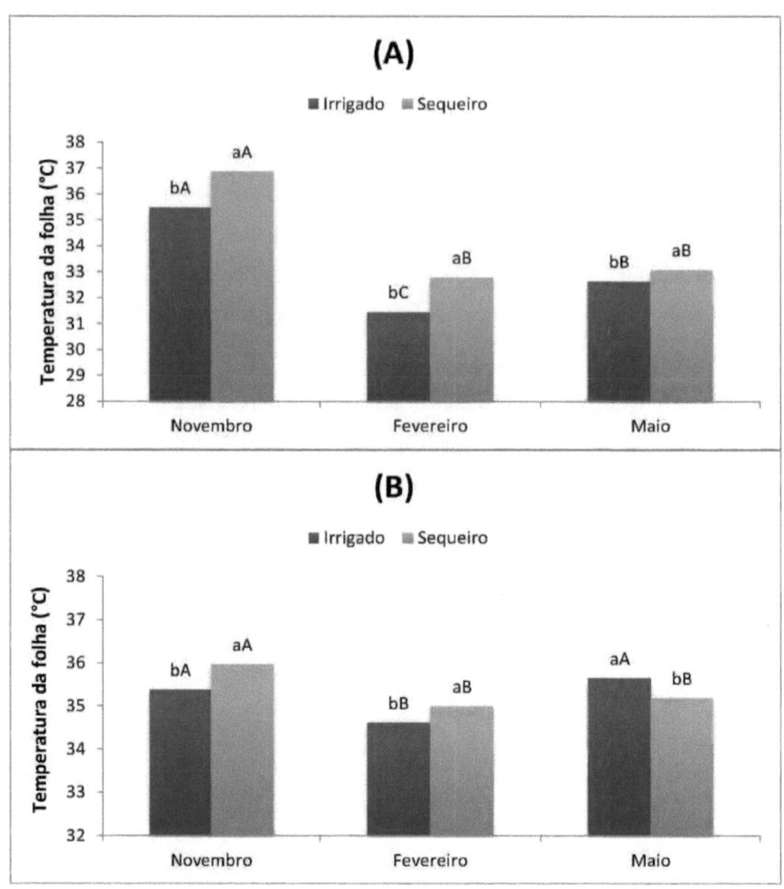

Averages followed by the same letter do not differ from each other at 5% probability, using the Tukey test. Lowercase letters (between water regime treatments) and uppercase letters (between seasons within the same water regime) Source: Author's elaboration.

Ludlow and Muchow (1990) apud Fiuza (2010) consider that keeping the leaf temperature equal to or slightly lower than the air temperature proves the cooling capacity of different species, via transpiration, in order to keep the plant protected from very high temperature ranges, and that this response is desirable and positive for the plant's adaptation to water stress. Some authors have pointed out the possible causes of the increase in leaf temperature in stressed plants, as well as showing Tf as a factor that modifies some physiological parameters in plants. Silva *et al.* (2004), working with ten Caatinga species at the beginning of the dry season, observed that most species' leaf temperature remained above that of the air, and associated the increase in leaf temperature with an increase in diffusive resistance in *Ziziphus joazeiro*, *Schinopsis brasiliensis* and *Bauhinia cheilantha* as a result of the decrease in available water content in the soil. Machado *et al.* (2005), in a study of *Citrus* varieties, concluded that leaf temperature affects the rate of CO_2 assimilation through effects on stomatal conductance and carboxylation efficiency. Dias and Marenco (2007) found a reduction in conductance (*gs*) as leaf temperature increased in *Swietenia macrophylla*, but there was no effect of this factor on photoinhibition, which is the slow and reversible reduction in photosynthesis as a result of exposure to full sunlight. Cordeiro *et al.* (2009) in a study of this same species correlated leaf temperature negatively with conductance (*gs*) and transpiration (*E*) in plants under water stress.

Figura 24 - Relationship between leaf temperature (T_{leaf}) and air temperature as a function of season in six tree species subjected to irrigated and rainfed conditions. Acaraù-CE, 2013.

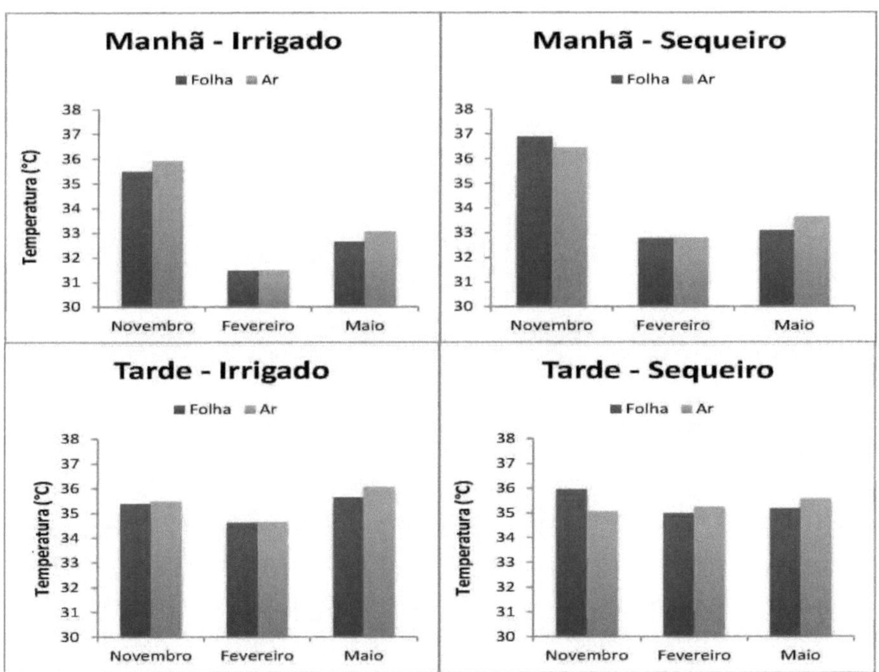

Source: Author's elaboration.

4.2.3 Photosynthetic efficiency of nitrogen (A/N) and phosphorus (A/P) use

Table 7 shows the summary of the analysis of variance for the photosynthetic efficiency of nitrogen (*A/N*) and phosphorus (*A/P*) variables. It was found that only the triple interaction between the factors did not influence *A/N*, while all the isolated factors and their interactions were significant for the *A/P* variable.

Table 7 - Summary of the analysis of variance for the variables photosynthetic efficiency of nitrogen

(*A/N)* and phosphorus (*A/P*) use in the afternoon in six tree species subjected to irrigated and rainfed conditions in November, February and May. Acaraù-CE, 2013.

F.V	G.L	Q.M	
		A/N	*A/P*
Species (a)	5	3042,0**	5.0×10^6 **
Waste - a	12	247,9	$2,2 \times 10^5$
Water regime (b)	1	13257,9**	8.2 x10 ^{*5}
a x b	5	1110,4*	1.8×10^6 **
Residue - b	12	220,76	$1,7 \times 10^5$
Season (c)	1	11060,8**	9.7×10^6 **
a x c	5	3227,6**	2.2×10^6 **
b x c	1	5012,1**	3.1×10^6 **
a x b x c	5	1172,7 $^{n.s}$	1.2 x 10 ^{*6}
Residue - c	24	460,7	$3,6 \times 10^5$
Total	71		
CV a (%)		17,76	18,39
CV b (%)		16,75	15,92
CV c (%)		24,20	23,42

G.L - degrees of freedom; (a) - species; (b) - water regime; (c) - season; CV - coefficient of variation; **, *, $^{n.s}$ - significant by the Tukey test at 1%, 5% and non-significant, respectively.

In relation to the photosynthetic efficiency of nitrogen use (*A/N*), it can be seen in Figure 25 that the seasonal rainfall altered the photosynthetic efficiency of nitrogen use only in the rainfed plants, showing a significantly higher *A/N* in the rainy season (May), with an increase of 76.27% compared to the drier period (November). Under these conditions these plants reduce the rate of CO_2 assimilation for each unit of mineral nitrogen contained in the leaves, but with the increase in water demand this efficiency is recovered. These results show that under these conditions these species are more responsive to the increase in *A/N* with seasonal changes due to the increase in CO_2 assimilation in the rainy season. As for the plants under irrigated conditions, they showed no significant change in *A/N* in relation to the periods, so this species makes better use of nitrogen for photosynthetic activity because they are in an environment where water is available in the soil.

Figura 25 - Photosynthetic efficiency of nitrogen use (*A/N)* in six tree species subjected to irrigated and rainfed conditions as a function of season. Acaraù-CE, 2013.

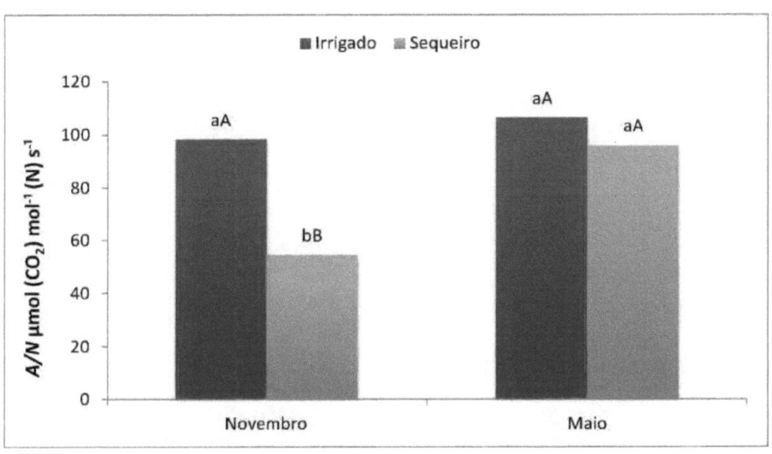

Averages followed by the same letter do not differ at 5% probability, using the Tukey test. Source: Prepared by the author.

Figure 26 shows that the photosynthetic efficiency of phosphorus use *A/P* was significantly altered by rainfall seasonality in marupà and ipê-rosa under irrigated conditions. For rainfed conditions, it was found that with the exception of gonçalo alves, all species had their *A/P* negatively affected by dry season conditions (November) with reductions of 116.9% in mahogany, 120.68% in guanandi, 45.54% in marupà, 163.13% in ipê-amarelo and 54,11% in ipê-rosa, which shows that these species have their efficiency of phosphorus use in photosynthesis negatively affected by water stress, however, they have a great capacity to recover when soil water availability increases.

The effect of water stress on the photosynthetic efficiency of nitrogen use can be attributed to effects on plant morphogenesis, i.e. growth and structure, especially of the leaf, as a result of damage caused at cellular level when these plants are subjected to conditions of low water availability. One of these effects is a reduction in specific leaf area. For plants under rainfed conditions, it was observed that the *A/N* was higher when the specific leaf area was greater, i.e. during the rainy season.

Figura 26 - Photosynthetic efficiency of phosphorus use (*A/P*) in six tree species subjected to irrigated and rainfed conditions as a function of season. Acaraù-CE, 2013.

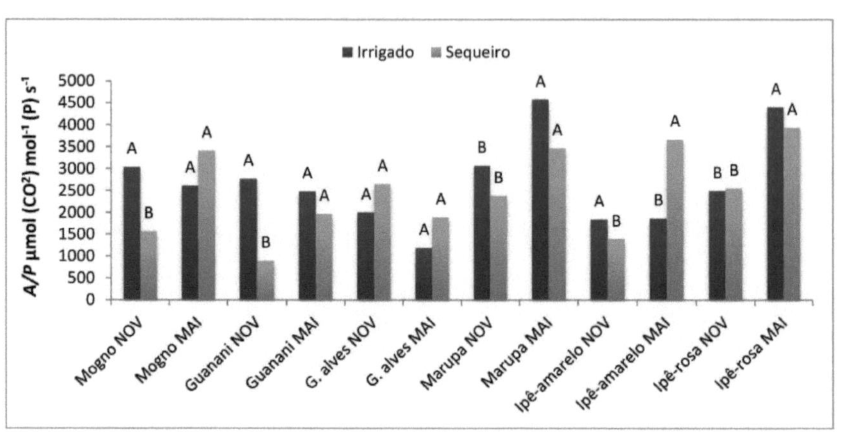

Averages followed by the same letter do not differ at 5% probability, using the Tukey test. Source: Prepared by the author.

Table 8 - Averages of nitrogen and phosphorus content and photosynthetic efficiency in the use of N and P in six tree species subjected to irrigated and rainfed conditions as a function of season. Acaraù-CE, 2013.

Species	N (g.Kg-)[1]		*A/N* (µmol co2 mol-i N s-i)		P (g.Kg-i)		*A/P* (µmol CO2 mol-i P s-i)	
	Nov	May	Nov	May	Nov	May	Nov	May
Mahogany I	19,32	21,53	95,58	70,93	1,34	1,29	3038,69	1570,32
Mahogany S	18,13	19,01	51,05	120,82	1,30	1,48	2610,68	3406,57
Guanandi I	13,19	13,65	82,27	82,49	0,86	1,00	2771,67	892,870
Guanandi S	15,76	13,60	25,85	56,72	1,00	0,86	2482,77	1970,40
G. Alves I	23,23	25,29	129,27	70,29	1,30	3,29	2007,08	2652,82
G. Alves S	24,21	33,28	77,33	77,00	1,56	2,97	1190,55	1898,57
Marupà I	20,19	15,45	89,41	121,39	1,29	0,90	3070,32	2385,38
Marupà S	17,67	17,20	60,39	82,96	0,98	0,90	4580,05	3471,81
Yellow Ipê I	21,95	22,10	111,60	139,35	2,92	3,62	1845,60	1393,77
Yellow Ipê S	20,66	32,92	43,75	104,24	1,43	2,06	1872,68	3667,48
Ipê-rosa I	20,35	17,25	81,04	153,33	1,45	1,32	2496,38	2554,79
Ipê-rosa S	17,31	24,78	67,82	133,31	1,01	1,85	4409,00	3937,30

I - Irrigated; S - Rainfed; Nov - November and May - May. Source: Prepared by the author.

These results corroborate Mendes and Marenco (2013) who found this relationship for ten native tree species in the Central Amazon and Poorter and Evans (1998) who state that species with a larger specific leaf area are more efficient at using nitrogen. The levels of nitrogen and phosphorus in the leaves, both in terms of mass and area, were not associated with higher *A/N* and *A/P* in most species, showing that the amount of nitrogen in the leaves does not influence its use for photosynthesis in these species (Table 8). According to Larcher (2006), plants with high photosynthetic efficiency obtain high CO_2 gains even if the nitrogen content in the leaves is moderate.

4.3. SPAD index

The SPAD index was found to be significantly influenced by the single factors species, water regime and season, as well as by the interaction species *vs* water regime and species *vs* season (Table 9).

Table 9 - Summary of the analysis of variance for the SPAD index in six tree species under irrigated and rainfed conditions as a function of season. Acaraù-CE, 2013.

F.V	G.L	Q. M
		SPAD index
Species (a)	5	3060,20**
Waste - a	24	7,47
Water regime (b)	1	1074,00**
a x b	5	42,89*
Residue - b	24	12,82
Season (c)	1	148,74**
a x c	5	36,97*
b x c	1	39,44 n.s
a x b x c	5	26,32 n.s
Residue - c	48	13,29
Total	119	
CV a (%)		5,14
CV b (%)		6,73
CV c (%)		6,85

G.L - degrees of freedom; (a) - species; (b) - water regime; (c) - season; CV - coefficient of variation; **, *, n.s - significant by the Tukey test at 1%, 5% and non-significant, respectively.

Figure 27-A shows that only ipê-rosa showed no significant difference in SPAD index in relation to the water regime, while all the other species showed lower values when subjected to dry conditions.

Figure 27 - SPAD index in six tree species as a function of water regime (A) and season (B). Acaraù-CE, 2013.

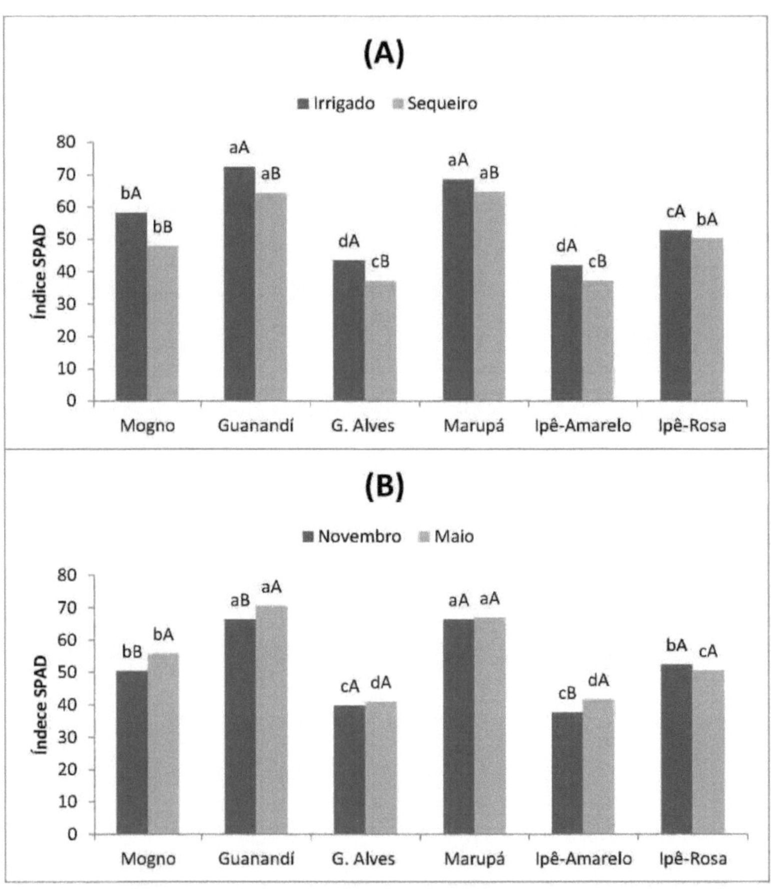

Averages followed by the same letter do not differ from each other at 5% probability, using the Tukey test. (A) upper case (between water regime treatments) and lower case (between species within treatments); (B) upper case (between period treatments) and lower case (between species within period treatments). Source: Author's elaboration.

The results show that ipê-rosa is able to maintain chlorophyll levels in the leaf even under rainfed conditions with reduced water availability. However, in the other species, reduced water availability caused negative effects on chlorophyll levels. According to Rey *et al.* (1999), water stress, at certain levels, favors the formation of reactive oxygen species, which damage photosynthetic pigments, among other cellular structures, which may explain the reduction in chlorophyll levels under these conditions.

Among the species under irrigated conditions, guanandi (72.35) and marupà (68.58) had the highest SPAD values, followed by mahogany (58.21), ipê-rosa (52.62), gonçalo alves (43.46) and ipê-amarelo (41.92). Considering the rainfed regime, marupà (64.64) and guanandi (64.31) obtained the highest averages, followed by ipê-rosa (50.23), mahogany (47.77), ipê-amarelo and gonçalo alves (37.10). The results showed that taking into account the interaction between species *and* water regime, water availability had a negative influence on the SPAD index, causing a decrease in the chlorophyll content in the leaf.

Figure 27-B shows the effect of the interaction between species and seasons on the SPAD index. Mahogany, guanandi and ipê-amarelo showed increases in leaf chlorophyll content in the rainy season (May) of 10.77%, 6.12% and 10.67%, respectively, compared to the dry season (November). Gonçalo alves, marupà and ipê- rosa did not significantly change their leaf chlorophyll levels depending on the season, so they were not influenced by the evaluation periods. With regard to the species, in the dry season it was found that guanandi (66.33) and marupà (66.33) had the highest SPAD averages, followed by ipê-rosa (52.33), mahogany (50.28), Gonçalo Alves (39.75) and ipê-amarelo (37.55). In the rainy season, guanandi (70.36) and marupà (66.92) had the highest averages, followed by mahogany (50.70), ipê-rosa (50.52), ipê-amarelo (41.56) and gonçalo alves (40.81).

The response observed for the relative chlorophyll content of the leaf in plants under water stress, estimated using the SPAD index, varies according to the species studied. Fernandes (2012), in a study of four eucalyptus clones, found no influence of water regimes (with and without irrigation) on SPAD values. Batista (2012) and Moura (2010), on *Jatropha curcas* plants, observed a reduction in the SPAD index in plants subjected to water deficiency and attributed this reduction to the lower relative water content in the leaves found in plants under water deficiency. Mendes *et al.* (2013) observed that *Cordia oncocalyx* plants in native Caatinga forest reduced their SPAD index values fifty days after the onset of drought. Water stress affects the biosynthesis of chlorophyll, which is why its content can vary significantly in plants subjected to water deficiency (CARVALHO *et al.*, 2007), and these same authors found changes in chlorophyll content in cerrado plants as a function of seasonality.

4.4. Chlorophyll fluorescence (Quantum efficiency of photosystem II)

With the exception of the single factor season, all the other factors and their interaction significantly influenced the chlorophyll fluorescence variable (*Fv/Fm*) (Table 10).

Table 10 - Summary of the analysis of variance for chlorophyll *a* fluorescence in six tree species under irrigated and rainfed conditions as a function of season. Acaraù-CE, 2013.

F.V	G.L -	Q.M
		Fv/Fm
Species (a)	5	0,029**
Waste - a	12	0,005
Water regime (b)	1	0,075**

a x b	5	0,011*
Residue - b	12	0,003
Season (c)	1	0,001 n.s
a x c	5	0,011**
b x c	1	0,012*
a x b x c	5	0,017**
Residue - c	24	0,002
Total	71	
		9,40
		7,60
CV a (%) CV b (%) CV c (%)		6,73

G.L - degrees of freedom; (a) - species; (b) - water regime; (c) - season; CV - coefficient of variation; **, *, n.s - significant by the Tukey test at 1%, 5% and non-significant, respectively.

Figure 28 shows that in all the irrigated species, the *Fv/Fm* did not change significantly depending on the evaluation period. This is partly due to the soil moisture conditions which did not cause any water stress to these plants during the dry season. For plants under dry conditions, gonçalo Alves, marupà and ipê-rosa did not significantly change their *Fv/Fm* depending on the period evaluated. Mahogany and guanandi had an increase in *Fv/Fm* of 22.12% and 48.82%, respectively, in the rainy season (May) compared to the dry season (November). The marupà showed a reduction of 15.62% in the rainy season compared to the dry season. The results show that, taking *Fv/Fm* into account, only mahogany and guanandi had the photochemical efficiency of the photosynthetic process negatively affected by the condition of hydric stress resulting from the seasonal variation in rainfall, with this efficiency being affected during the dry period, which may have occurred due to damage to the photosynthetic apparatus of these species in such a condition and characterized the effect of photoinhibition. According to Maxwell and Johnson (2000), *Fv/Fm* values lower than 0.75 indicate that the plant is under stress and that the quantum efficiency of photosystem II has been reduced. In this study, lower values were only found in mahogany (0.541) and guanandi (0.555) under dry season conditions. According to Taiz and Zeiger (2013), values between 0.75 and 0.85 indicate the efficiency of the leaf photosynthetic apparatus in converting light energy into FSII. In this experiment, all the plants under irrigated conditions in the two evaluation periods showed values within this range, demonstrating efficiency in this parameter.

Figure 28 - Chlorophyll *a* fluorescence in six tree species subjected to irrigated and rainfed conditions as a function of season. Acaraù-CE, 2013.

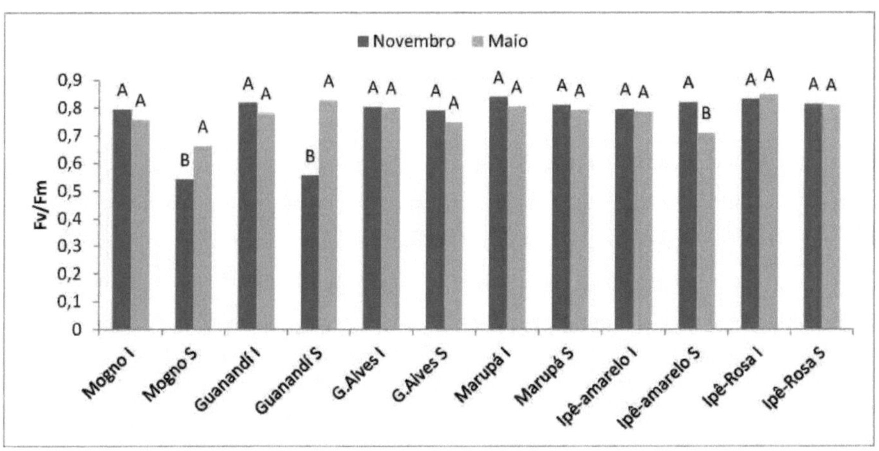

Averages followed by the same letter do not differ at 5% probability, using the Tukey test. I - Irrigated; S - Rainfed. Source: Prepared by the author.

Fv/Fm is considered to be the maximum efficiency at which the light absorbed by FSII is used for the reduction of Q_A(quinone A), which is the primary FSII acceptor related to the use of light energy for the reduction of $NADP^+$, indispensable for carbon assimilation. Lower *Fv/Fm* values indicate a reduction in the efficiency of excitation energy capture by the open reaction centers of the FSII, indicating photochemical energy dissipation (BAKER, 2008). Changes in *Fv/Fm* in plants subjected to water stress have been reported by several authors. Trovao *et al.* (2007), evaluating tree species in the Caatinga during the dry and rainy seasons, found that of the eleven species, seven did not change the *Fv/Fm* in relation to the periods and the other four showed minimal reductions and attributed this to the evolutionary characteristics of these species. Esposti (2013) observed a reduction in *Fv/Fm* in *Handroanthus chrysotrichus* 15 days after the start of water restriction, while in *Cariniana estrellensis* this parameter was not altered under the same conditions. Lemos Filho (2000), working with three species from the Cerrado, observed that the species *Annona crassifólia* did not change its *Fv/Fm* in the dry and rainy seasons, unlike *Eugenia dysenterica* and *Campomanesia adamantium*, which showed higher *Fv/Fm* in the rainy season. Gonçalves *et al.* (2009) observed a 23% reduction in *Fv/Fm* after 14 days of irrigation suppression in young *Carapa guianensis* plants and attributed this to the reduction in photochemical activity due to the decrease in *gs* caused by water stress and reported that photoinhibition occurs under more severe drought conditions. In this study, this situation was only observed for mahogany and guanandi under water stress.

4.5. Specific leaf area and degree of leaf succulence

There was a significant effect of the single factors species and season and of the interactions between the single factors for the specific leaf area variable. For this variable, the single factor hydric regime was not significant. The degree of leaf succulence was significantly influenced by the species, water regime and season and their interactions, except for the water regime *vs.* season interaction (Table 11).

Figure 29 shows that the specific leaf area (SFA) of mahogany, guanandi, gonçalo alves and marupà was practically unaffected between periods under irrigated conditions, but ipê-amarelo and ipê-rosa showed a significant increase of 32.47% and 39.18% in the rainy season (May) under the same conditions. Mahogany, gonçalo alves, ipê-amarelo and ipê-rosa had lower AFE in the dry season (November) with averages of 0.92 $dm^2 g^{-1}$, 1.00 $dm^2 g^{-1}$, 0.99 $dm^2 g^{-1}$ and 0.68 dm2 g ι respectively. Guanandi and marupà showed no significant difference in EFA in relation to the two periods, with averages of 0.67 dm2 g-i and 0.70 dm2 g-i in the dry period and (0.57 dm2 g i) and (0.72 dm2 g i) in the rainy period, respectively.

Table 11 - Summary of the analysis of variance for the variables of specific leaf area (SLA) and degree of leaf succulence (GS) in six tree species subjected to irrigated and rainfed conditions as a function of season. Acaraù-CE, 2013.

F.V	G.L -	QM	
		AFE	GS
Species (a)	5	1,00**	5,248**
Waste - a	18	0,01	0,015
Water regime (b)	1	0,05 n.s	0,068**
a x b	5	0,06*	0,078**
Residue - b	18	0,02	0,007
Season (c)	1	1,17**	0,550**
a x c	5	0,24**	0,068**
b x c	1	0,20**	0,022n.s
a x b x c	5	0,06**	0,070**
Residue - c	36	0,01	0,017
Total	95		
CV a (%)		11,70	7,00
CV b (%)		15,40	4,56
CV c (%)		12,08	7,32

G.L - degrees of freedom; (a) - species; (b) - water regime; (c) - season; CV - coefficient of variation; **, *, n.s - significant by the Tukey test at 1%, 5% and non-significant, respectively.

Figura 29 - Specific leaf area of six tree species subjected to irrigated and rainfed conditions as a function of season. Acaraù-CE, 2013.

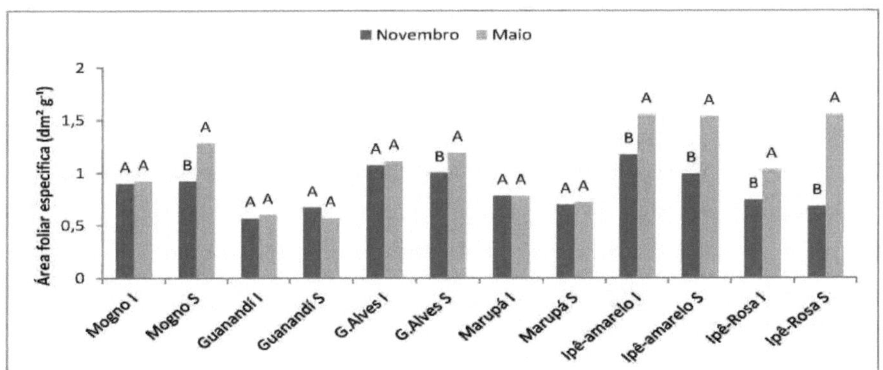

Averages followed by the same letter do not differ at 5% probability, using the Tukey test. I - Irrigated; S - Rainfed. Source: Prepared by the author.

The lower EFA seen in plants subjected to water stress during the dry season can be understood as a plant strategy to reduce the transpiration surface by reducing the area per unit mass, as a way of saving water during periods of scarcity. According to Larcher (2006), leaves that develop under conditions of deficient water supply are generally smaller and have a smaller specific leaf area. These results are different from those found by Nascimento *et al.* (2011) who found no significant differences in the SFA *of Hymenaea courbaril* L. seedlings when subjected to different soil water contents. Araujo and Haridasan (2007) found no significant differences in EFA in sixteen cerrado species depending on the dry and rainy seasons. In view of the results, it can be concluded that the reduction in EFA in plants under water deficiency varies according to the species as a result of the long evolutionary period in a given region.

Figure 30 shows the degree of leaf succulence (GS) of the species in the different water regimes as a function of season. The plants under irrigated conditions showed a higher GS in the dry period (November) with values of 2.76 g $_{H2O}$ dm$^{\wedge 2}$ in guanandi, 1.29 g $_{H2O}$ dm^2 in gonçalo Alves, 2.12 g $_{H2O}$ dm^2 in marupà, 1.28 g $_{H2O}$ dm^2 in ipê-amarelo and 2,05 g $_{H2O}$ dm$^-$ 2 in the ipê-rosa, but only the mahogany showed a significant reduction in GS as a function of the periods, obtaining greater GS in the dry period (1.78 g $_{H2O}$ dm$^-$ 2) in relation to the rainy period (November) (1.51 g $_{H2O}$ dm$^-$ 2). In the dryland plants, it was found that mahogany, guanandi, marupà and ipê-rosa had higher GS values in the dry season, with values of 1.59 g $_{H2O}$ dm$^-$ 2,2.93 g $_{H2O}$ dm$^-$ 2, 2,10 g $_{H2O}$ dm$^-$ 2 and 2.08 g $_{H2O}$ dm$^-$ 2 respectively. This difference was only significant for mahogany, guanandi and ipê-rosa, with reductions of 23.25%, 10.56% and 38.66% respectively. For gonçalo alves and ipê-amarelo, increases of 5.26% and 3.81% were observed in the rainy season compared to the dry season, although these were not significant.

The fact that plants subjected to water stress show higher GS during periods of lower water availability can be seen as a strategy to maintain leaf hydration by storing water, protecting the plant from sudden wilting and leaf shrinkage. In this study, this strategy was not observed only in gonçalo alves and ipê-amarelo, a fact that may be associated with the characteristics of the leaves of these species, which are thinner (higher AFE) than the other species with average values of 1.09 dm2 g-i and 1.31 dm2 g-i, which decreases this ability to store water in drier periods. On the other hand, plants with thicker leaves (lower AFE) such as guanandi, marupà and ipê-rosa had higher GS values and a greater capacity to store water in the leaves. According to Larcher (2006) a special form of water conservation is the use of carbohydrates capable of hydrating themselves (mucilage) in cells, in ducts and in cavities between cells.

Figura 30 - Degree of leaf succulence (GS) in six tree species subjected to irrigated and rainfed conditions as a function of season. Acaraù-CE, 2013.

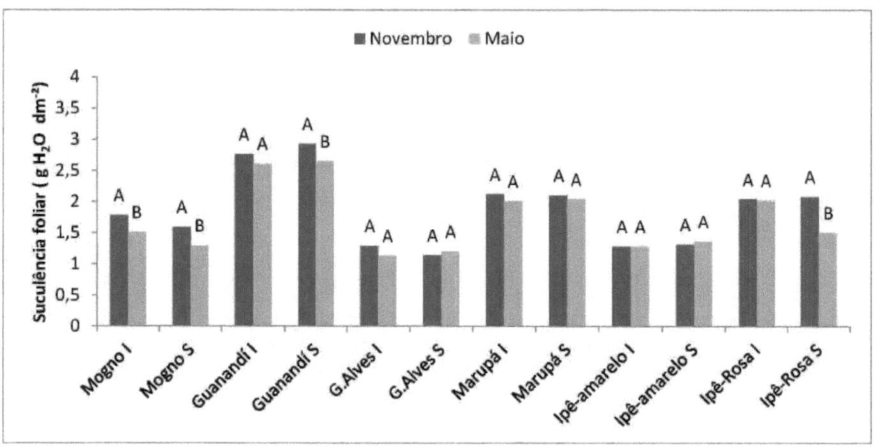

Averages followed by the same letter do not differ at 5% probability, using the Tukey test. I - Irrigated; S - Rainfed. Source: Author's elaboration.

4.6. Growth analysis

4.6.1. Plant height and stem diameter

Table 12 shows that, with the exception of the single factor water regime, all the single factors and interactions significantly influenced the plant height variable. For the stem diameter variable, it was found that the single factor water regime and the double interaction water regime *vs.* time had no significant effect. For the other single factors and interactions, the effect was significant.

Figure 31 shows the height values of the six species in the two water regimes as a function of time. It can be seen that after 12 months, with the suspension of irrigation in the rainfed treatment, mahogany, guanandi and gonçalo alves had lower plant height values in subsequent months as a result of the water stress. At the end of 24 months, this difference was 36.41%, 44.16% and 47.63% greater in irrigated plants in mahogany, guanandi and gonçalo alves, respectively. In the marupà ipê-amarelo and ipê-rosa plants, the height of the plants in the water stress treatment remained higher than in the irrigated plants, in which case the first year of irrigation was sufficient for the establishment of these species, with heights greater than or equal to the irrigated conditions even after the water supply was reduced. The heights were 20.81% and 27.62% greater in the rainfed treatment at the end of 24 months for marupà and yellow ipê, respectively. In the ipê-rosa, the reduction was minimal in the irrigated plants, around 4.63% at the end of the two-year period.

Table 12 - Summary of the analysis of variance for the variables plant height and stem diameter for six tree species subjected to irrigated and rainfed conditions as a function of time. Acaraù-CE, 2013.

F.V	G.L -	Q. M	
		HEIGHT	DAP
Species (a)	5	107,19**	53,59**
Waste - a	18	0,36	0,99
Water regime (b)	1	$0,01^{n.s}$	$0,19^{n.s}$
a x b	5	4,94**	7,01**
Residue - b	18	0,17	0,42
Time (c)	4	132,45**	232,24**
a x c	20	3,01**	9,32**
b x c	4	0,67**	$0,42^{n.s}$
a x b x c	20	0,56**	1,22**
Residue - c	144	0,05	0,13
Total	239		
CV a (%)		28,77	41,36
CV b (%)		19,60	26,94
CV c (%)		11,58	15,32

G.L - degrees of freedom; (a) - species; (b) - water regime; (c) - time; CV - coefficient of variation; **, *, n.s - significant by the Tukey test at 1%, 5% and non-significant, respectively.

Figure 32 shows the stem diameter values of the six species in the two water regimes as a function of time. It can be seen that mahogany, gonçalo alves and ipê-rosa had lower stem diameter values than irrigated plants after irrigation was suspended, with reductions of 54.09%, 36.52% and 17.83%, respectively, at the end of 24 months. In ipê-rosa, this reduction was more significant in the last half of the period. In the case of guanandi, marupà and ipê-amarelo, the plants maintained their higher stem diameter values in the rainfed treatment, which shows that these species were already well established in relation to the water supply. At the end of the two-year period, the diameter values were 27.71%, 28.13% and 30.88% higher in the dryland regime for guanandi, ipê- amarelo and marupà, respectively.

Figure 31 - Plant height of Mahogany (*Swietenia macrophylla* King), Guanandi (*Calophyllum brasiliense* Cambess.), Gonçalo Alves (*Astronium fraxinifolium*), Marupà (*Simarouba amara* Aubl.), Ipê-Amarelo (*Tabebuia serratifolia* (Vahl.) Nich.) and Ipê-Rosa (*Tabebuia impetiginosa*), subjected to irrigated and rainfed conditions as a function of time. Acaraù-CE, 2013.

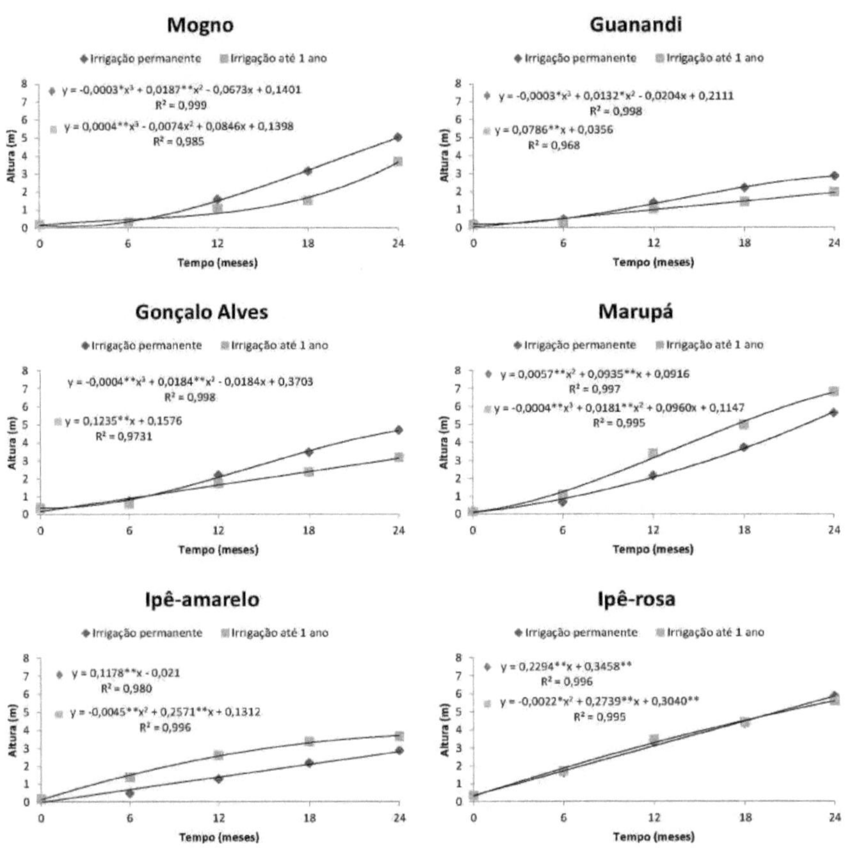

Source: Author's elaboration.

The main effects of water stress on plant growth are associated with a decrease in turgidity and the limitation of metabolism, mainly in the synthesis of proteins and amino acids. The reduction in the synthesis of protein metabolism causes cell division to be interrupted, reducing the speed of the mitotic process and thus reducing the growth process, particularly growth extension (LARCHER, 2006).

Figure 32 - Stem diameter of Mahogany (*Swietenia macrophylla* King), Guanandi (*Calophyllum brasiliense* Cambess.), Gonçalo Alves (*Astronium fraxinifolium*), Marupà (*Simarouba amara* Aubl.), Ipê-Amarelo (*Tabebuia serratifolia* (Vahl.) Nich.) and Ipê-Rosa (*Tabebuia impetiginosa*), subjected to irrigated and rainfed conditions as a function of time. Acaraù-CE, 2013.

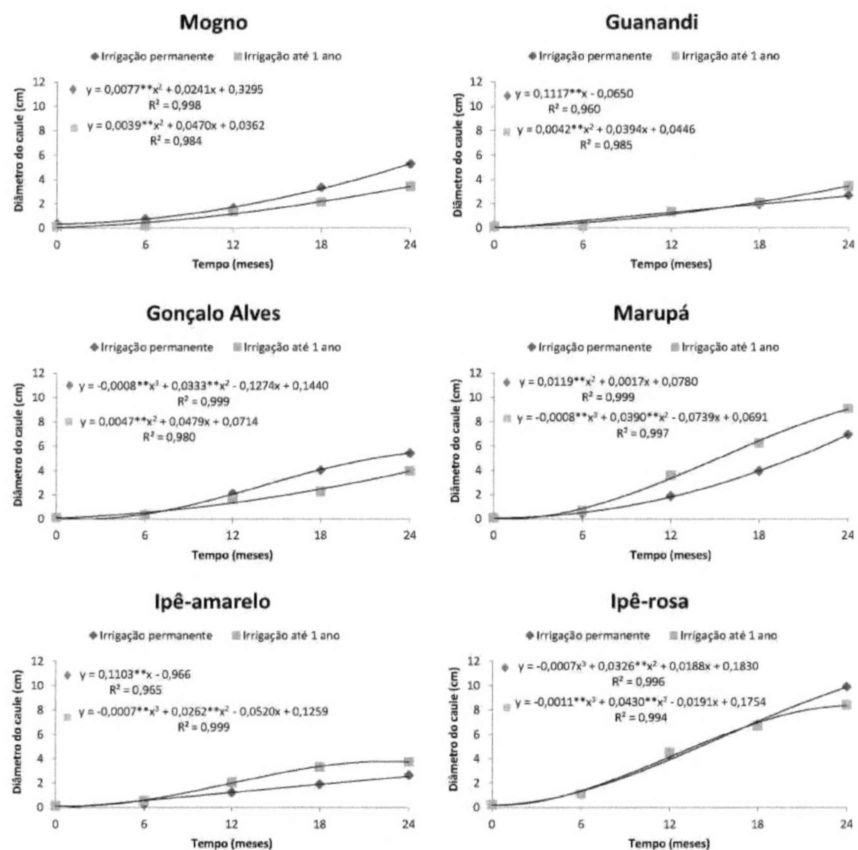

Fonte: Elaboração do autor.

In this study, the greatest reductions in height growth due to water deficiency were seen in gonçalo alves, guanandi and mahogany. For stem diameter, the greatest reductions were seen in alves and mahogany. These reductions may be related to the reductions in the rate of CO_2 assimilation seen in these plants when they were subjected to water stress as a result of stomatal closure in response to the water *status of* the soil (Figure 12, Figure 13 and Figure 14). This reduction in the rate of assimilation has a direct effect on the production of photoassimilates, which is reflected in the plant's growth. It was found that the species ipê-amarelo, ipê-rosa and marupà in dry conditions maintained a pattern of height and stem diameter equal to or greater than irrigated plants over the 24-month period. The resistance of these plants to conditions of low water availability in the soil may be associated with various adaptations developed by these plants to tolerate these conditions. With regard to mechanisms related to gas exchange, but specifically stomatal closure, it was found that during drier periods the plants reduce stomatal conductance as a way of saving water (Figure 14). As for water use efficiency, it was observed that the plants had better *A/E* and *A/gs* in the drier periods (Figure 18 and Figure 20). In relation to the reduction in the specific leaf area as a strategy for reducing the area of transpiration, only ipê- amarelo and ipê-rosa reduced their EFA during periods of lower rainfall (Figure 29).

It is important to add that other strategies used by plants to adapt to and survive drought that were not evaluated in this work should be taken into account, such as the allocation of dry mass in the root by increasing the root to aerial part ratio, as well as the growth of the deep root zone and the fact that the soil in the experimental area favors this growth, as it is deep and sandy in texture as discussed above, this parameter should be given important relevance to the growth data obtained.

4.6.2. Absolute growth rate (AGR) and relative growth rate (RGR)

Table 13 shows that all the isolated factors and their interactions significantly influenced the variables absolute growth rate in height (ATR (H)), relative growth rate in height (RGR (H)) and diameter (RGR (D)). For the variable absolute growth rate in diameter (ATR (D)), it was found that the single factor water regime had no significant effect. For the other single factors and interactions, the effect was significant.

Figure 33 shows the absolute growth rate in height of the six species in the two water regimes as a function of time. There is a drop in the absolute growth rate in height (TCA) of the rainfed plants after irrigation was suspended in the 12th month for all species. However, with the exception of ipê-amarelo, the growth rate increased again after the 18th month as a result of the rainy season. For most of the species, the growth rates best fitted a cubic model resulting from fluctuations in the water supply; the yellow ipe showed a decreasing linear behavior under rainfed conditions. Under irrigated conditions, mahogany, guanandi, marupà and ipê-amarelo showed absolute growth rates in height adjusted to a quadratic model tending to decrease at the end of the two-year period. ipê-rosa and gonçalo alves showed greater oscillation with water availability tending to decrease in the 18th month followed by an increase in the 24th month.

Table 13 - Summary of the analysis of variance for the variables absolute growth rate in height and diameter and relative growth rate in height and diameter for six tree species subjected to irrigated and rainfed conditions as a function of time. Acaraù-CE, 2013.

F.V	G.L	Mean squares

		TCA (H)	TCA (D)	TCR (H)	TCR (D)
Species (a)	5	0,1126**	0,3624**	0,0125**	0,0125**
Waste - a	18	0,0017	0,0063	0,0002	0,0003
Water regime (b)	1	0,0080**	0,0006 n.s	0,0006*	0,0039**
a x b	5	0,0167**	0,0369**	0,0009**	0,0014**
Residue - b	18	0,0008	0,0031	0,0001	0,0001
Season (c)	3	0,1447**	0,5603**	0,2061**	0,3720**
a x c	15	0,0282**	0,0335**	0,0235**	0,0151**
b x c	3	0,0215**	0,0276**	0,0084**	0,0078**
a x b x c	15	0,0067**	0,0171**	0,0068**	0,0080**
Residue - c	108	0,0024	0,0020	0,0014	0,0008
Total	191				
CV a (%)		24,44	36,52	12,39	12,32
CV b (%)		16,99	25,54	9,72	8,95
CV c (%)		29,16	20,78	31,13	20,73

G.L - degrees of freedom; (a) - species; (b) - water regime; (c) - time; CV - coefficient of variation; **, *, n.s - significant by the Tukey test at 1%, 5% and non-significant, respectively.

Figure 34 shows the absolute growth rates in diameter of the species. It was found that under irrigated conditions mahogany and marupà showed linear growth, while ipê-rosa, ipê-amarelo and guanandi suffered fluctuations in growth rates with a cubic behavior. Gonçalo alves best fitted a quadratic model with a growth peak in the 18th month.

Under rainfed conditions, there was a drop in absolute growth rates in diameter in all species. The drop in ACR (D) was less pronounced in marupà. In ipê-amarelo and ipê-rosa, TCA (D) tended to decrease over time until the end of the 24-month period just after irrigation was stopped. In mahogany, guanandi and gonçalo alves, the ACR (D) decreased after the water supply was interrupted, but increased from the 18th month onwards when the rainy season began.

Figure 33 - Absolute growth rate in height (ATR (H)) in Mahogany (*Swietenia macrophylla* King), Guanandi (*Calophyllum brasiliense* Cambess.), Gonçalo Alves (*Astronium fraxinifolium*), Marupà (*Simarouba amara* Aubl.), Ipê-Amarelo (*Tabebuia serratifolia* (Vahl.) Nich.) and Ipê-Rosa (*Tabebuia impetiginosa*) under irrigated and rainfed conditions as a function of time. Acaraù-CE, 2013.

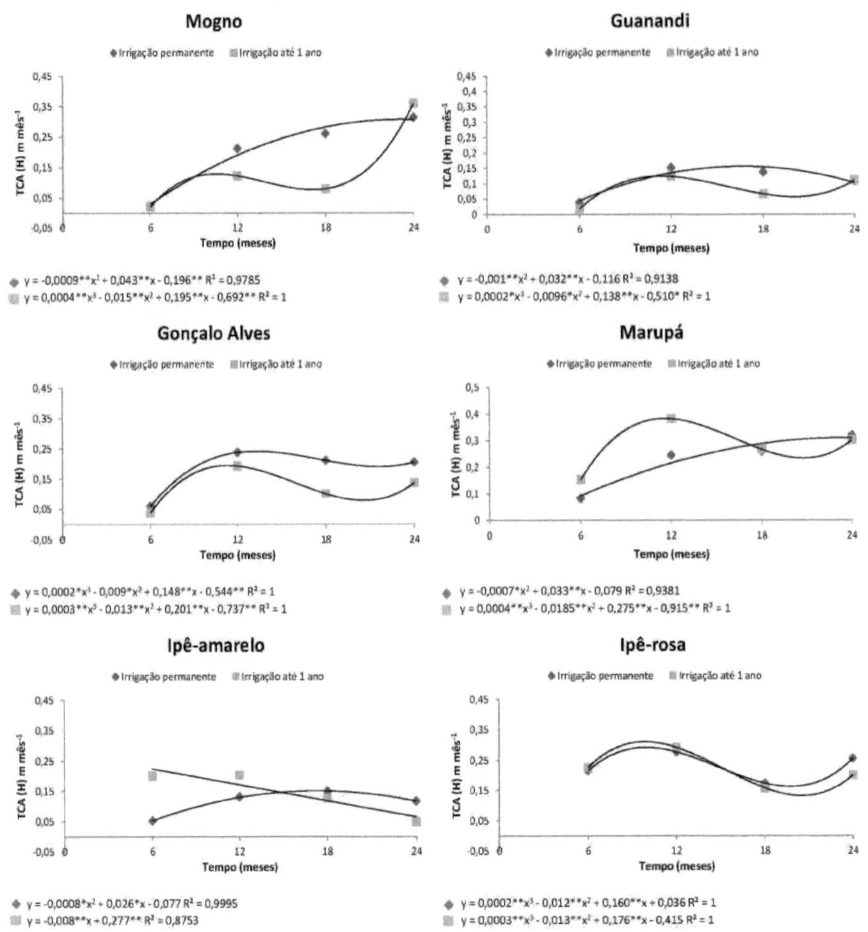

Source: Author's elaboration.

Figure 34 - Absolute diameter growth rate (ADR (D)) in Mahogany (*Swietenia macrophylla* King), Guanandi (*Calophyllum brasiliense* Cambess.), Gonçalo Alves (*Astronium fraxinifolium*), Marupà (*Simarouba amara* Aubl.), Ipê-Amarelo (*Tabebuia serratifolia* (Vahl.) Nich.) and Ipê-Rosa (*Tabebuia impetiginosa*) under irrigated and rainfed conditions as a function of time. Acaraù-CE, 2013.

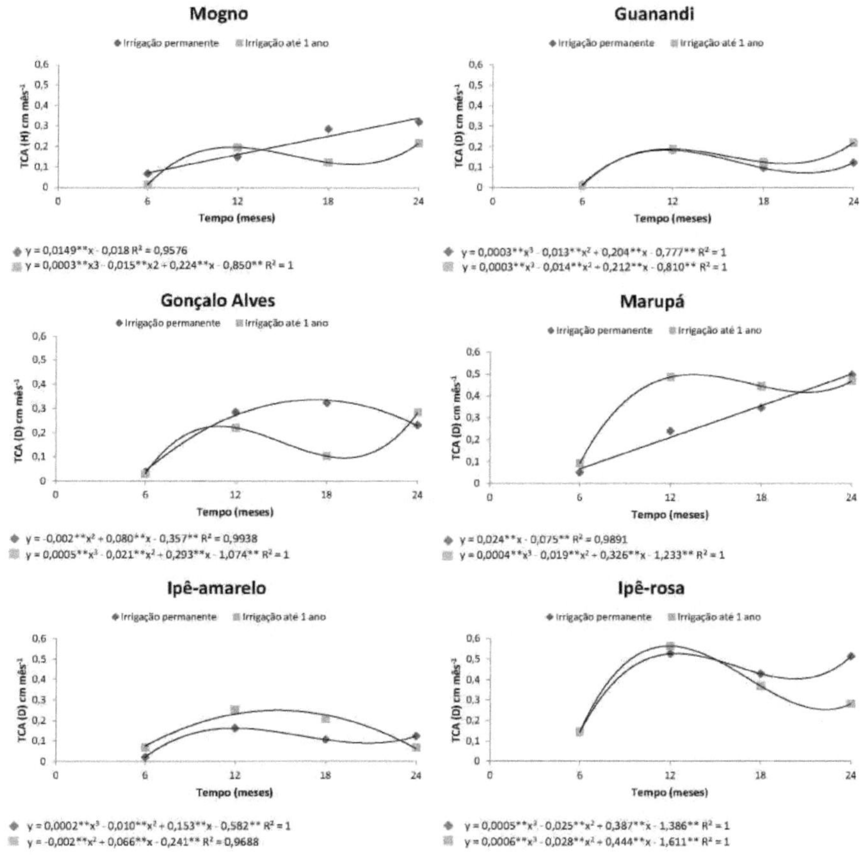

Source: Author's elaboration.

Figure 35 shows the relative height growth rate of the six species. It was found that in mahogany, guanadi, gonçalo alves and ipê-rosa the RCT (H) varied similarly in the two water conditions over time in relation to the alternation of seasons and water supply, however the RCT (H) was higher in the permanently irrigated plants. Ipe-yellow and marupà showed differences in TCR (H) trends, with linear behavior in irrigated plants and quadratic behavior in rainfed plants, both decreasing as a function of time and with higher values in irrigated conditions.

Figure 35 - Relative height growth rate (RHT (H)) in Mahogany (*Swietenia macrophylla* King), Guanandi (*Calophyllum brasiliense* Cambess.), Gonçalo Alves (*Astronium fraxinifolium*), Marupà (*Simarouba amara* Aubl.), Ipê-Amarelo (*Tabebuia serratifolia* (Vahl.) Nich.) and Ipê-Rosa (*Tabebuia impetiginosa*) under irrigated and rainfed conditions as a function of time. Acaraù-CE, 2013.

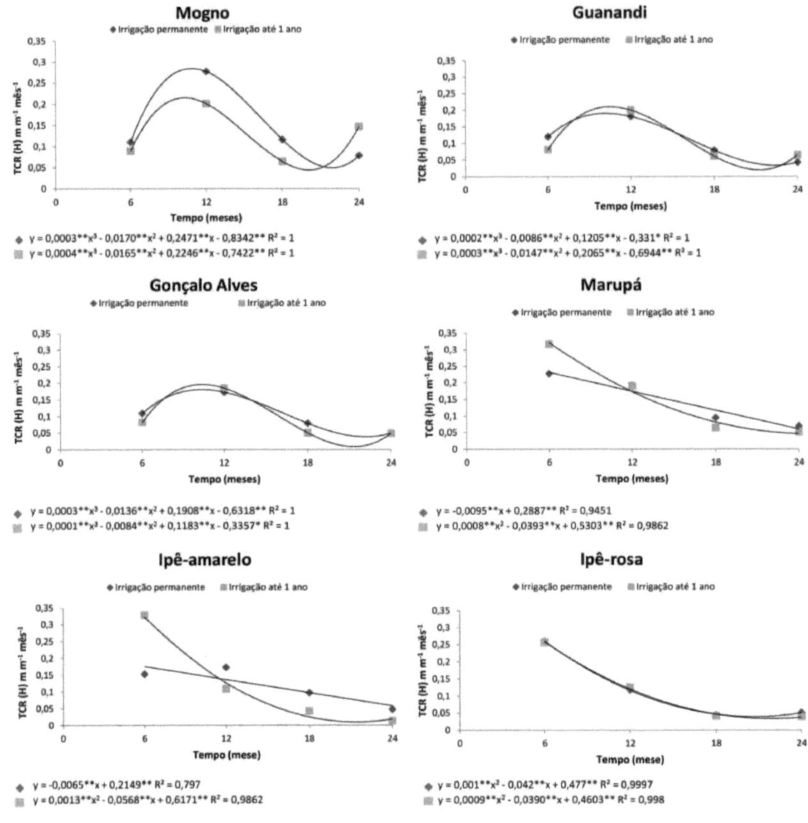

Source: Author's elaboration.

With regard to the relative growth rate in diameter of the species, Figure 36 shows a very sharp downward trend in TCR (D) in all species except mahogany, under both water regime conditions. However, it was slightly more effective under rainfed conditions. Under irrigated conditions, mahogany showed minimal variation in TCR (D) with a decreasing linear behavior.

Figure 36 - Relative growth rate in diameter (RGR (D)) in Mahogany (*Swietenia macrophylla* King), Guanandi (*Calophyllum brasiliense* Cambess.), Gonçalo Alves (*Astronium fraxinifolium*), Marupà (*Simarouba amara* Aubl.), Ipê-Amarelo (*Tabebuia serratifolia* (Vahl.) and Ipê-Rosa (*Tabebuia impetiginosa*) under irrigated and rainfed conditions as a function of time. Acaraù-CE, 2013.

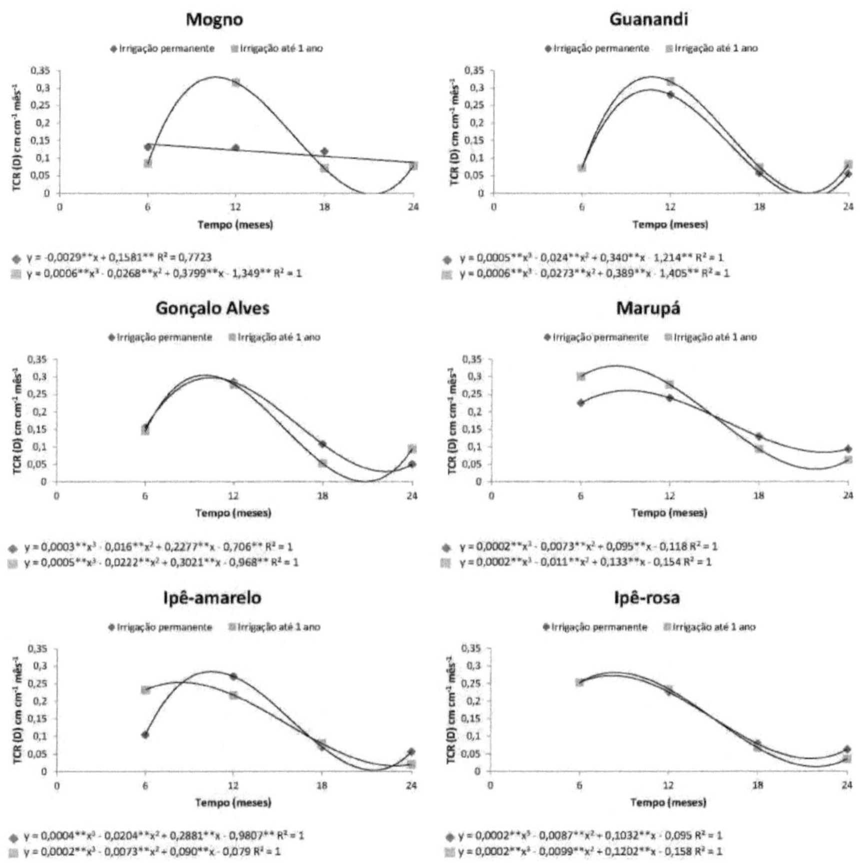

Source: Author's elaboration.

The first year of irrigation was sufficient for the establishment of some species, which resulted in a smaller drop in growth after the suspension of the water supply, so that they were not exposed to stress in the initial period of development, a critical point for the growth of the species. In this sense, the time of planting is extremely important for the management of dryland plantations. The drop in TCA (H) six months after suspension of the water supply was 43.60% in marupâ, 54.53% in mahogany, 61.11% in ipê-amarelo, 85.98% in ipê-rosa, 86.36% in guanandi and 91.08% in gonçalo Alves. For the ACD variable (D), the drop in rates over the same period was 9.45% in marupà, 20.67% in ipê-amarelo, 51.20% in guanandi, 52.84% in ipê-rosa, 60.65% in mahogany and 113.59% in gonçalo alves. At 12 months after the cessation of irrigation, with the exception of ipê-amarelo, all the species showed increases in TCA (H) as a result of the rainy season, the most significant being mahogany with an increase of 355.69%, for TCA (D) the greatest increases were observed in gonçalo alves with 176.69% and in mahogany 77.04%. When the same analysis was carried out six months after the start of water restriction for relative growth rates, it was found that TCR (H) fell by 159.52% in ipê-amarelo, 196.87% in marupà, 204.87% in ipê-rosa, 214.06% in mahogany and 270% in gonçalo alves. For TCR (D), the decline was 174.68% in ipê-amarelo, 197.84% in marupà, 249.25% in ipê-rosa, 329.72% in guanandi, 345.07% in mahogany and 424.52% in gonçalo alves. After 12 months, there was a tendency for TCR (H) to increase, but only significantly in mahogany, with an increase of 129.68%. For the same period, the increase in TCR (D) was more significant in gonçalo alves, with 77.35%, and in mahogany, with 11.26%.

Seasonal variation in water availability is one of the main factors influencing plant productivity in the tropics. Variations in physiological characteristics are very marked in the different seasons and tend to be reflected in vegetative growth rates in plants and consequently in wood production. It was found that under rainfed water conditions, growth rates follow rainfall variations over the two-year period in all six species, although this is more pronounced in some species. The availability of water in the soil can be considered a limiting factor, but the variability of other climatic factors in the seasons is very considerable, with the vapor pressure deficit being more prominent in the dry seasons, interfering in soil-plant-atmosphere water relations and influencing mechanisms that regulate growth. Variations in growth rates are also related to the different phenological phases, which occur in different ways and at different speeds depending on the species. Declines in absolute and relative growth rates are closely linked to the oscillations observed in CO_2 assimilation rates, caused mainly by the availability of water in the soil. These responses to water deficit conditions depending on the season can be evidenced by stomatal closure. According to Larcher (2006), the main effects of a reduction in stomatal conductance are a reduction in the production of photoassimilates followed by an increase in the activity of oxidizing enzymes which is directly reflected in a reduction in plant growth. Studying the long-term growth dynamics of tree plants, Locoselli *et al* (2012), using the chronology of tree ring widths, observed that in the species *Hymenaea courbaril* and *Hymenaea Stigonocarpa*, climatic variations, especially rainfall, influenced plant growth in the different forest strata. In a similar study with *Mimosa acantholoba* in a tropical forest in Mexico, Brienen *et al* (2010) found a close relationship between the increase in growth and the rainy season and added that in years with low rainfall, growth was reduced by 37%. Anholetto Junior (2013) analyzed the growth of *Cedrela odorata* and found that climatic seasonality had a direct influence on tree trunk growth in both the Caatinga and the Atlantic Forest in the state of Sergipe. Brienen and Zuidema (2005) in a study of six species in a tropical forest in Bolivia also found a positive relationship between tree growth and periods of rain. Ferreira (2002) in a study of 13 forest species found that rainfall and temperature are directly correlated with tree circumference growth. Lima *et* al (2010), working with *Tynanthus cognatus* Miers, also verified the influence of rainfall on the growth of this species. Knowledge of growth fluctuations in different species is of the utmost importance for planning management strategies for forest species, especially in regions where rainfall events are difficult to predict, such as semi-arid or subsumed dry regions like the irrigated perimeter of Baixo Acaraù.

5. CONCLUSIONS

1. The water deficit caused by seasonal rainfall induces reductions in the rate of photosynthesis in the six species studied under rainfed conditions.

2. Stomatal conductance is reduced in the drier seasons in plants subjected to rainfed conditions.

3. Transpiration is reduced in the drier seasons in plants subjected to rainfed conditions.

4. Plants under conditions of water restriction increase the intrinsic and momentary efficiency of water use in the driest period.

5. In dry seasons, irrigated plants control leaf temperatures better, keeping them lower than the air temperature.

6. The internal/external CO_2 ratio was negatively affected in mahogany, guanandi and ipê- amarelo during the driest season.

7. The photosynthetic efficiency of nitrogen use is reduced by water deficit.

8. The photosynthetic efficiency of phosphorus use is negatively affected by water restriction during the dry season in rainfed plants.

9. Low water availability in the soil reduces the quantum efficiency of photosystem II in mahogany and guanandi.

8. The SPAD index is reduced by water deficit in the dry season in mahogany, guanandi, gonçalo alves, ipê-amarelo and marupâ. When considering seasons, mahogany, guanandi and ipê- amarelo show reductions in the dry period.

9. Water stress caused by seasonal rainfall decreases the specific leaf area of mahogany, gonçalo alves, marupâ, ipê-amarelo and ipê-rosa under dryland conditions. The degree of leaf succulence increases in the dry season in mahogany, guanandi, marupâ and ipê-rosa under rainfed conditions.

10. The ipê-amarelo, ipê-rosa and marupâ species irrigated for up to 1 year maintained a higher pattern of height and stem diameter than the permanently irrigated plants over the 24-month period.

11. Water stress causes falls in absolute and relative growth rates in height and diameter in the six species studied.

12. Growth rates follow the seasonal variation of rainfall in rainfed plants.

6. REFERENCES

ABRAF - Brazilian Association of Planted Forest Producers. **ABRAF statistical yearbook 2012 base year 2011**. Brasilia, 2012, 150 p.

ALBUQUERQUE, M. P. F.; MORAES, F. K. C.; SANTOS, R. I. N.; CASTRO, G. L. S.; RAMOS, E. M. L. S.; PINHEIRO, H. A. Ecophysiology of Young African Mahogany Plants Subjected to Water Deficit and Rehydration. **Pesquisa Agropecuâria. Brasileira**, Brasilia, v. 48, n. 1, p. 9-16, 2013.

ALMEIDA, A.C., SOARES, J.V. Comparison of water use in Eucalyptus grandis plantations and dense ombrophilous forest (Atlantic Forest) on the east coast of Brazil. **Revista Arvore**, v.27, n.2, p. 159 - 170. 2003.

ALMEIDA, L. P.; ALVARENGA A. A.; CASTRO A. M.; ZANELA, S. M.; VIEIRA, C. V. Initial Growth of Plants *of Cryptocaria aschersoniana* Mez. Subjected to Levels of Solar Radiation. **Ciência Rural**, Santa Maria, v. 34, n. 1, p. 83-88, 2004.

ANDREASSIAN, V. Waters and forests: from historical controversy to scientific debate. **Journal of Hydrology**, v. 291, p. 1-27, 2004.

ANGELOCCI, L. R. **Water in the plant and gas/energy exchange with the atmosphere:** introduction to biophysical treatment. Piracicaba: Author's edition. 2002. 272 p.

ANHOLETTO JUNIOR, Claudio Roberto. **Dendroecology and isotopic composition (^{13} C) of the growth rings of *Cedrela odorata* trees, Meliaceae, in the Caatinga and Atlantic Forest in the State of Sergipe**. 2013. 90 f. Dissertation (Master's Degree) - Escola Superior de Agricultura "Luiz de Queiroz", Piracicaba, 2013.

ARAÙJO, J. F.; HARIDASAN, M. Relationship between deciduousness and leaf nutrient concentrations in woody species of the cerrado. **Revista Brasileira.de Botânica**, V.30, n.3, p.533-542, 2007.

ATROCH, E.M.A.C. (2008) **Effects of abiotic factors on growth, photosynthetic characteristics and synthesis of volatile oils in young plants of lauraceae species in Central Amazonia**. Thesis (Doctorate in Botany), Manaus-AM, Instituto Nacional de Pesquisas da Amazônia/Universidade Federal do Amazonas, 109 p.

BAKER, N. R. Chlorophyll fluorescence: A probe of photosynthesis *in vivo*. **Annual Review of Plant Biology**, v. 59, p. 89 - 113, 2008.

BARROSO, G. M.; MORIM, M. P.; PEIXOTO, A. L.; ICHASO, C. L. **F. Frutos e sementes: morfologia aplicada à sistemàtica de dicotiledôneas**. Viçosa - MG: Editora UFV, 1999. 444p.

BARUCH, Z. Responses to drought and flooding in tropical forage grasses. 2. Leaf water potential, photosynthesis rate and alcohol dehydrogenase activity. **Plant and Soil**, v. 164, n. 1, p. 97-105, 1994.

BASTOS, A. de M. **As madeira do Parâ**. Arquivos do Serviço Florestal, Rio de Janeiro, v. 2, n. 2, p. 157-182, 1946.

BATISTA, Lourdes Regina Lopes. **Root and physiological characteristics of jatropha (Jatropha curcas L.) propagated by seed and cuttings grown under different water conditions**. 2012. 47 f. Dissertation (Master's Degree in Agronomy: Plant Production) - Federal University of Alagoas, Rio Largo, 2012.

BENITEZ RAMOS, R. F.; MONTESINOS LAGOS, J. L. **Catalogo de ciem especies forestales de Honduras: distribución, propriedades y usos**. Siguatepeque: Escuela Nacional de Ciencias Forestales, 1988. 200p.

BLACKMAM, F. F. Optima and limiting factors. **Ann. Bot**, n.19, p 281-295, 1905.

BRAZIL. Secretariat for Strategic Affairs. **Guidelines for structuring a national policy on planted forests**. Brasilia, 2011. 104 p.

BRIENEN, R. J. W.; LEBRIJA-TREJOS; E.; ZUIDEMA, P. A.; MARTINEZ-RAMOS, M. Climate-growth analysis for a Mexican dry forest tree shows strong impact of sea surface temperatures and predicts future growth declines. **Global Change Biology**. v. 16, p. 2001-12. 2010.

BRIENEN, R. W.; ZUIDEMA, P. A. Relating tree growth to rainfall in Bolivian rain forests: a test for six species using tree ring analysis. **Oecologia**, Berlin, v.146, p. 1-12. 2005.

BRITO, J. O.; BARRICHELO, L. E. G. **Considerações sobre a produçâo de carvão vegetal com madeira da Amazônia**. Serie Técnica IPEF, Piracicaba, v.2, n. 5, p, 1-27, 1981.

BROWN, A. E. *et al.* A review of paired catchment studies for determining change in water yield resulting from alterations in vegetation. **Journal of Hydrology**, v. 310, p. 28-61, 2005.

BRUCE, W. B.; EDMEADES, G. O.; BARKER, T. C. Molecular and physiological approaches to maize improvement for drought tolerance. **Journal of Experimental Botany**, v.53, n.66, p.13-25, 2002.

BRUIJNZEEL, L. A. Hydrological functions of tropical forests: not seeing the soil for the trees? **Agriculture, Ecosystems and Environment**, v. 104, p. 185-228, 2004.

CABRAL, E. L.; BARBOSA, D. C. de A.; SIMABUKURO, E. A. Growth of young plants of *Tabebuia aurea* (Manso) Beth. & Hook. f. ex S. Moore submitted to hydric stress. **Acta Botànica Brasileira**, v. 18, n. 2, p. 241-251, 2004.

CALDATO, S. L.; SCHUMACHER, M. V. Water use by forest plantations - a review. **Ciência Florestal**, Santa Maria, v.23, n.3, p.507-516, 2013.

CAMPELLO, F. B.; GARIGLIO, M. A.; SILVA, J. A.; LEAL, A. M. A. **Diagnóstico Florestal da Regiao Nordeste**. Brasilia: IBAMA, 1999. 20p.

CAMPOSTRINE, E. **Chlorophyll a fluorescence: Theoretical considerations and practical applications.** State University of Northern Fluminense. Handout, 2001.

CARVALHO, A. P. F.; BUSTAMANTE, M. M. C.; KOZOVITS, G. P. Seasonal variations in pigment and nutrient concentrations in leaves of cerrado species with different phenological strategies. **Revista Brasileira de Botànica**, Sao Paulo, v. 30, n. 1, p. 19-27, 2007.

CARVALHO, P. E. R. **Espécies Arbóreas Brasileiras.** Brasilia, DF: Embrapa Informaçao Tecnológica; Colombo: Embrapa Florestas, 2003. v. 1, 1039 p.

CARVALHO, P. E. R. **Espécies Arbóreas Brasileiras.** Brasilia, DF: Embrapa Informaçao Tecnológica; Colombo: Embrapa Florestas, 2006. v. 2, 627 p.

CARVALHO, P. E. R. **Espécies Arbóreas Brasileiras.** Brasilia, DF: Embrapa Informaçao Tecnológica; Colombo: Embrapa Florestas, 2008. v. 3, 593 p.

CARVALHO, P. E. R. **Espécies Arbóreas Brasileiras.** Brasilia, DF: Embrapa Informaçao Tecnológica; Colombo: Embrapa Florestas, 2010. v. 4, 644 p.

CAVALCANTE, M. S. **Biological deterioration and preservation of wood**. Sao Paulo: IPT, 1982. 40 p.

CEARA. State Superintendence for the Environment. **State Forestry Program - PEF**. Fortaleza, 2004. 29 p.

CERNUSAK, L. A. *et al.* Large variation in whole-plant water-use efficiency among tropical tree species. **New Phytologist**, Malden, v. 173, n. 2 p.294-305, 2007.

CHAVES, J. H. *et al.* Early selection of eucalyptus clones for environments with different soil water availability: water relations of plants in tubes. **Revista Arvore**, v.28, n.3, p.333-341, 2004.

CHAVES, M. M.; PEREIRA, J. S.; MAROCO, J.; RODRIGUES, M. L.; RICARDO, P.P.; SÓRIO, M. L.; CARVALHO, I.; FARIA, T.; PINHEIRO, C. How plants cope with water stress in the field. Photosynthesis and growth. **Annals of Botany**, v.89, p.907-916, 2002.

CODOGNOTTO, L. M.; SANTOS, D. M. M.; LEITE, I. C.; MARIN, A.; MADALENO, L. L.; KOBORI, N. N.; BANZATTO, D. A. Effect of aluminum on the chlorophyll content of mung bean and labe-labe seedlings. **Ecossistema**, Espirito Santo do Pinhal, v. 27, n. 1, p. 27-30, 2002.

CORDEIRO, Y. E. M.; PINHEIRO, H. A.; SANTOS FILHO, B. G.; CORRÊA, S. S.; SILVA, J. R. R.; DIAS-FILHO, M. B. Physiological and morphological responses of young mahogany (*Swietenia macrophylla* King) plants to drought. **Forest Ecology and Management**, v.258, p.1449-1455, 2009.

CORDEIRO, Y.E.M. **Potential for use in the recovery of degraded areas: A study of three native species from Eastern Amazonia under two water regimes**. Dissertation. Federal University of Parà, Belém, 2012. 89 p.

CORREIA, K. G.; FERNANDES, P. D.; GHEYI H. R.; NOBRE, R. G.; SANTOS, T. D. Growth, production and chlorophyll a fluorescence characteristics in peanut under salinity conditions. **Revista Ciência Agonômica**, Fortaleza, v. 40, n. 4, p. 414-521, 2009.

CORREIA, K.G.; NOGUEIRA, R.J.M. C. Evaluation of the growth of peanut (*Arachis hypogaea L.*) submitted to water deficit. **Revista de Biologia e Ciências da Terra**, Belo Horizonte, v.4, n.2, 2004.

COSTA, G.F.; MARENCO, R.A. Photosynthesis, stomatal conductance and leaf water potential in young andiroba trees (*Carapaguianensis*). **Acta Amazonica**, v.37, p.229-234, 2007.

DELGADO, Luiz Gustavo Martinelli. **Production of native seedlings under different water managements**. 2012. 96 f Dissertation (Master's Degree in Forestry Sciences) - Universidade Estadual Paulista, Faculdade de Ciências Agronômicas, Botucatu, 2012.

NATIONAL DEPARTMENT OF WORKS AGAINST DROUGHTS - DNOCS. Executive Coordination Groups for Agricultural Operations (GOA). Situation on April 30, 1991. Fortaleza, 1991. NATIONAL DEPARTMENT OF WORKS AGAINST DROUGHTS - DNOCS.

Perimetroirrigadobaixo-acaraù . Fortaleza2012 . Available at:

http://www.dnocs.gov.br/~dnocs/doc/canais/perimetros_irrigados/ce/baixo_acarau.html.

Accessed on April 8, 2013.

DIAS, D. P.; MARENCO, R. A. Photosynthesis and photoinhibition in mahogany and acariquara as a function of luminosity and leaf temperature. **Pesquisa Agropecuària Brasileira**, Brasilia, v.42, n.3, p.305-311, 2007.

DONAGEMA, G.K.; CAMPOS, D.V.B. de; CALDERANO, S.B.; TEIXEIRA, W.G.; VIANA, J.H.M. (Org.). **Manual of soil analysis methods**. 2.ed. rev. Rio de Janeiro: Embrapa Solos, 2011. 230p.

BRAZILIAN AGRICULTURAL RESEARCH CORPORATION (EMBRAPA). National Soil Research Center. **Brazilian Soil Classification System**. Brasilia: Embrapa Produçao de Informaçao,

Rio de Janeiro, 2013, 3ª ed. 353p.

BRAZILIAN AGRICULTURAL RESEARCH COMPANY - EMBRAPA. **Manual of chemical analysis of soils, plants and fertilizers**. 2nd revised and expanded edition. Brasilia, DF: Embrapa Informaçao Tecnológica, 2009. 627 p.

ESPOSTI, Maria Stela de Oliveira Degli. **Water stress in two tree species of different successional stages**. 2013. 46 f. Dissertation (Master's Degree in Ecology and Natural Resources) - Universidade Estadual Do Norte Fluminense, Campos Dos Goytacazes - RJ, 2013.

FACCO, A. G.; RIBEIRO, A.; PRUSKI F. F.; MONTEIRO W. C.; LEITE, F. P.; ANDRADE. R. G.; MENEZES, S. J. M. C. Geo-information techniques for estimating water balance in eucalyptus. **Pesquisa Agropecuària Brasileira**, v.47, n.9, p. 1243-1250, 2012.

FARIAS, C. H. A. *et al.* Water use efficiency in sugarcane under different irrigation rates and zinc levels on the Paraiba coast. **Engenharia. Agricola**, Jaboticabal, v. 28, n. 3, p. 494-506, 2008.

FERNANDES, Emanuel Tâssio. **Photosynthesis and initial growth of Eucalyptus clones under different water regimes**. 2012. 113 f. Dissertation (Master's Degree in Agronomy/Phytotechnics) - State University of Southwest Bahia, Vitória da Conquista - BA, 2012.

FERREIRA, D. F. SISVAR: a program for statistical analysis and teaching. **Revista Symposium**, v.6, p.36-41, 2008.

FERREIRA, L.; CHALUB, D.; MUXFELDT, R. **Ipê-amarelo: *Tabebuia serratifolia* (Vahl) Nichols**. Technical Newsletter Amazon Seed Network, Manaus, v. 5, 2004.

FERREIRA, Ligia. **Periodicity of growth and wood formation of some tree species in semi-deciduous seasonal forests in the southeastern region of the state of São Paulo**. 2002. 103 f. Dissertation (Master's Degree) - Escola Superior de Agricultura "Luiz de Queiroz", Piracicaba, 2002.

FIUZA, Danila da Silva. **Identification of agronomic and physiological characteristics related to drought tolerance in cassava.** 2010. 61 f. Dissertation (Master's Degree in Agricultural Sciences/Phytotechnics) - Federal University of Recôncavo da Bahia, Cruz das Almas-BA, 2010.

FLEXAS, J.; MEDRANO, H. Drought-inhibition of photosynthesis in C3 plants: stomatal and non-stomatal limitations revisited. **Annals of Botany**, v.89, p.183-189, 2002.

FYLLAS, N. M. et al. Basin-wide variations in foliar properties of Amazonian forest: phylogeny, soils and climate. **Biogeosciences**, v.6, n.4, p.2677-2708, 2009.

GARRIDO, A. O. **Silvicultural characteristics and nutrient content in the leaf litter of some pure and mixed stands of native species.** 1981. 105 f. Dissertation (Master's Degree in Forestry Sciences)

- Escola Superior de Agricultura "Luiz de Queiroz", Piracicaba, 1981.

GONÇALVES, E. R.; FERREIRA, V. M.; SILVA J. V,; ENDRES, L.; BARBOSA, T. B.; DUARTE, W. G. Gas exchange and chlorophyll a fluorescence in sugarcane varieties submitted to water deficiency. **Rev. Bras. De Eng. Agricola**, Campina Grande, v. 14, n.4, p. 378-386, 2010.

GONÇALVES, J. F. C.; SILVA, C. E. M.; GUIMARÂES, D. G. Photosynthesis and leaf water potential of young andiroba plants submitted to water deficiency and rehydration. **Pesquisa. Agropecuâria. Brasileira**, Brasilia, v.44, n.1, p.8-14, 2009.

GRISI, F. A.; ALVES, J. D.; CASTRO, E. M. de; OLIVEIRA, C. de; BIAGIOTTI, G.; MELO, L. A. de. Leaf anatomical evaluations of 'catuai' and 'siriema' coffee seedlings subjected to water stress. **Ciência e Agrotecnologia**, Lavras, v.32, n.6, p.1730-1736, Nov./Dec., 2008.

HOEL, B.O., SOLHAUG, K.A. Effect of irradiance on chlorophyll estimation with the Minolta SP AD-502 leaf chlorophyll meter. **Annals of Botany** - London, v.82, p. 389-392, 1989.

HUBER, A. J.; TRECAMAN, R. V. Water use efficiency in Pinus radiata plantations in Chile. **Bosque**, v. 25, n. 3, p. 33-43, 2004.

ILSTEDT, U. *et al.* The effect of afforestation on water infiltration in the tropics: A systematic review and meta-analysis. **Forest Ecology and Management**, v. 251, p. 45-51, 2007.

INSTITUTE OF FOREST RESEARCH AND STUDIES. **Elaboration of a list of native tree species for silviculture and multiple-use models**. Available at:< http://www.ipef.br/pcsn/documentos/especies_nativas_silvicultura.pdf>. Accessed on July 16, 2013.

JANKOWSKY, i. P.; CHiMELO, J. P.; CAVALCANTE, A. de A.; GALiNA, i. C. M.; NAGAMURA, J. C. s. **Madeiras brasileiras**. Caxias do sul: spectrum, 1990. v. 1, 172 p.

JAROSZ, N. *et al.* Carbon dioxide and energy flux partitioning between the understorey and the overstorey of a maritime pine forest during a year with reduced soil water availability. **Agricultural and Forest Meteorology**, v. 148, p. 1508-1523, 2008.

JOKER, D.; SALAZAR, R. ***Calophyllum brasiliense*** **Cambess**. Humlebaek: Danida Forest Seed Centre, 2000. 2p. (Seed Leaflet, 46).

KENNAN, R.J. et al. Plantations and Water: Plantation Impacts on Stream Flow. Science for Decision Makers. Australian Government. Department of Agriculture, Fisheries and Forestry. Bureau of Rural Sciences . 2006. 8p . Available at : http://www.acera.unimelb.edu.au/materials/ brochures/SDM-PlantationsWater. Accessed on: April 10, 2013.

KRAUSE, G.H.; WEIS, E. Chlorophyll fluorescence and photosynthesis: the basics. Annual Review of Plant Physiology and Plant Molecular Biology, **Boca Raton**, v.42, p.313-349, 1991.

LARCHER, W. **Plant Ecophysiology**. Sao Carlos: RIMA, 2006. 531p.

LAWLOR, D. W.; CORNIC, G. Photosynthetic carbon assimilation and associated metabolism in relation to water deficits in higher plants. **Plant, Cell and Environment**, v. 25, n. 02, p. 275-279, 2002.

LECHINOSKI, A. Influence of water stress on the levels of protein and total soluble amino acids in teak leaves (*Tectona grandis* L. f.). **Revista Brasileira de Biociências**, Porto Alegre. v.5, supl.2, p.927-929, jul. 2007.

LEMOS FILHO, J. P. L. Photoinhibition in three cerrado species (*Annona crassifolia, Eugenia dysenterica* and *Campomanesia adamantium*) during the dry and rainy seasons. **Revista Brasileira de Botânica**, Sao Paulo, V.23, n.1, p.45-50, 2000.

LIBERATO, M. A. R.; GONÇALVES.; J. F. C.; CHEVREUIL, L. R.; NINA JUNIOR, A. R. N.; FERNANDES, A. V.; SANTOS JUNIOR, U. M. S. Leaf water potential, gas exchange and chlorophyll a fluorescence in acariquara seedlings (Minquartia guianensis Aubl.) under water stress and recovery. **Brazilian Journal Plant Physiology**, v. 18, n. 2, p. 315-323, 2006.

LIMA, A. C. ; PACE, R. M.; ANGYALOSSY. V. Seasonality and growth rings in lianas of Bignoniaceae. **Trees**, v. 24, p. 1045-1060, sep. 2010.

LIMA, D. de A.; ROCHA, M. G. Preliminary observations on the Mata do Buraquinho, Joao Pessoa, Paraiba. **Anais do Instituto de Ciências Biológicas**, Recife, v. 1, n. 1, p. 47-61, 1971.

LIMA, M. A.; BEZERRA, M. A.; FILHO, E. G.; PINTO, C. M.; FILHO, J. E. Gas Exchange in Sunny and Shaded Leaves of Cashew Tree under Different Water Regimes. Ver. **Revista Ciência Agronômica**, Fortaleza, v. 41, n. 4, p. 654-663, 2010.

LOCOSSELLI, G. M.; BUCKERIDGE, M. S.; MOREIRA, M. Z.; CECCANTINI, G. A multi-proxy dendroecological analysis of two tropical species *(Hymenaea* spp., Leguminosae) growing in a vegetation mosaic. **Trees**, v.27, n.1, p. 25-36, aug, 2012.

LOPEZ, J.A.; L1TILE JUNIOR, E.L.: RITZ, G.F.; ROMBOLD, J.S.; HAHN, W.J. **Arboles comunes del: Paraguay: riande yvyra mata kuera**. Washington: Cuerpo de Paz, 1987. 425p.

LORENZI, H. **Arvores brasileiras**: manual de identificaçao e cultivo de plantas arbóreas do Brasil. Nova Odessa: Plantarum, Sao Paulo, 2002, v. 1, 2. ed. 368p.

LUDLOW, M. M; MUCHOW, R. C. A critical evaluation of traits for improving crop yields in water-limited environments. **Advance in Agronomy**, San Diego, v.43, p.107-153, 1990.

MACHADO, E. C.; SCHIMIDT, P. T.; MEDINA, C. L.; RIBEIRO, R. V. Responses of photosynthesis of three citrus species to environmental factors. **Pesquisa Agropecuâria. Brasileira**, Brasilia, v.40, n.12, p.1161-1170, 2005.

MAINIERI, C.; CHIMELO, J. P. **Fichas de caracteristicas das madeira brasileiras**. Sao Paulo: IPT, 1989. 418p.

MARQUES, Marcia Cristina Mendes. **Autoecological study of guanandi (*Calophyllum brasiliense* Camb. Clusiaceae) in a riparian forest in the municipality of Brotas**, SP. 1994. 123f. . Dissertation (Master's Degree in Biology), State University of Campinas, SP, 1994.

MARRENCO, R.; GONÇALVES, J. F. C.; VIEIRA, G. Leaf gas exchange and carbohydrates in tropical trees differing in successional status in two light environments in central Amazonia. **Tree Physiology**, Victoria, Canada, v. 21, p. 1311-1318, 2001.

MAXWELL C.; JOHNSON, G. M. Chlorophyll fluorescence - a practical guide. **Journal of Experimental Botany**, Oxford, v. 51, n. 354, p. 659-668, 2000.

MEDRI, M.E. **Studies on waterlogging tolerance in native tree species from the Tibagi river basin**. Londrina: [s.n.], 2002. 172p.

MENDES, K. R.; MARENCO, R. A. Leaf traits and gas exchange in saplings of native tree species in the Central Amazon. **Scientia. Agricola**, Piracicaba, v.67, n.6, p.624-632, 2010.

MENDES, K. R.; MARENCO, R. A.; MAGALHÂES, N. S. Growth and photosynthetic efficiency of nitrogen and phosphorus use in Amazonian forest species in the juvenile phase. **Revista Arvore**, Viçosa-MG, v.37, n.4, p.707-716, 2013.

MENDES, M. M. S.; LACERDA, C. F.; FERNANDES, F. E. P.; CAVALCANTE, A. C. R.; OLIVEIRA, T. S. Ecophysiology of deciduous plants grown at different densities in the semiarid region of Brazil. **Theoretical and Experimental Plant Physiology**, v.25, n.2, p.97108, 2013.

MOUGET, J.; TREMBLIN, G. Suitability of the fluorescence monitoring system (FM, Hansatech) for measurement of photosynthetic characteristics in algae. **Aquatic Botany**, Amsterdam, v.74, p.219-231, 2002.

MOURA, Adelina Ribeiro de. **Morphological, physiological and biochemical aspects of *jatropha* (*Jatropha curcas* L.) subjected to water deficit**. 2010. 81 f. Dissertation (Master's Degree in Forest Science) - Federal Rural University of Pernambuco, Recife, 2010.

NASCIMENTO, H. H. C.; NOGUEIRA, R. J. M. C.; SILVA, E. C.; SILVA, M. A. Growth analysis of jatobà (*hymenaea courbaril* l.) seedlings in different levels of water in the soil. **Revista Arvore**, Viçosa, v.35, n.3, p.617-626, 2011.

NETO, M, T, C. Effect of water deficit on transpiration and stomatal resistance of mango. **Revista Brasileira de Friticiltira**, Jaboticabal, v. 25, n. 1, p. 93-95, 2003.

NINA JUNIOR, Adamir da Rocha. **Photosynthetic, nutritional and hydric characteristics of tree species in different forest strata of a primary forest in the Upper Rio Negro.** 2009. 88 f. Dissertation (Master's Degree in Tropical Forest Sciences) - National Institute for Amazonian Research, Manaus, 2009.

NOGUEIRA, A., MARTINEZ, C.A., FERREIRA, L.L. & PRADO, C.H.B.A. Photosynthesis and water use efficiency in twenty tropical tree species of differing succession status in a Brazilian reforestation. **Photosynthetica**, Heidelberg, v.42, n.3, p. 351-356, 2004.

NOGUEIRA, R. J. M. C. *et al.* **Ecophysiological aspects of drought tolerance in caatinga plants**. In: NOGUEIRA, R. J. M. C. *et al.* Estresses ambientais: danos e beneficios em plantas. Recife: Federal Rural University of Pernambuco, 2005.500p.

NOGUEIRA, R. J. M. C.; SILVA, E. C. Stomatal behavior in young plants of Schinopsis brasiliensis Engl. grown under water stress. **Iheringia**: Série Botànica, v. 57, n. 1, p. 31-38, 2002.

NOGUEIRA, R.J.M.C.; SANTOS, C.R.; NETO, E.B.; SANTOS, V.F. Physiological behavior of two peanut cultivars to different water regimes. **Pesquisa Agropecuària Brasileira**, v.33, p.1963-1969, 1998.

OKI, V.K., 2002. **Impacts of Pinus taeda harvesting on water balance, water quality and nutrient cycling in micro-watersheds**. Master's dissertation. ESALQ/USP, Piracicaba. 71p.

PAES, J. B.; MORAIS, V. M.; LIMA, C R. Natural resistance of nine woods from the Brazilian semi-arid region to fungi that cause soft rot. **Revista Arvore**, Viçosa, vol.29, n.3, pp. 365-371, 2005.

PAHLICH, E. Larcher's definition of plant stress: a valuable principle for metabolic adaptibility research. **Revista Brasileira de Fisiologia Vegetal**, Campinas, v.5, p.209-216. 1993.

PALHARES, D.; FRANCO, A. C.; ZAIDAN, L. B. P. Photosynthetic responses of cerrado plants in the dry and rainy seasons. **Revista brasileira de Biociências**, Porto Alegre, v. 8, n. 2, p. 213-220, 2010.

PALLARDY, S. G. **Physiology of wood plants**. 3ª ed. United States: Elsevier. 2008. 454 p.

PAULA, J. E.;ALVES, J. L. H. **Native woods: anatomy, dendrology, dendrometry, production, use**. Brasilia; Mokiti Okada Foundation, 1997. 541p.

PENNINGTON, T. D. **Meliaceae**. New York: New York Botanical Garden, 1981. 470 p.

PIMENTEL, C. **The relationship between plants and water.** Seropédica: EDUR, 2004. 191p.

PIMENTEL, C. **Carbon metabolism in tropical agriculture**. Seropédica: EDUR, 1998. 150 p.

PINCELLI, Renata Passos. **Tolerance to water deficiency in sugarcane cultivars assessed through morphophysiological variables**. 2010. 78p. Dissertation

(Faculty of Agronomic Sciences, UNESP - Botucatu Campus). Júlio de Mesquita Filho State University. Botucatu - SP, 2010.

PINZÓN-TORRES, J. A.; SCHIAVINATO, M. A. Growth, photosynthetic efficiency and water use efficiency in four tropical leguminous tree species. **Hoehnea,** v. 35, n. 3, p. 395-404, 2008.

POORTER, H.; EVANS, J. R. Photosynthetic nitrogen-use efficiency of species that differ inherently in specific leaf area. **Oecologia**, v. 116, n. 1, p. 26-37, 1998.

PORRA, R. J.; THOMPSON, W. A.; KRIEDEMANN, P. E. Determination of accurate extinction coefficients and simultaneous equations for assaying chlorophylls a and b extracted with four different solvents: verification of the concentration of chlorophyll standards by atomic absorption spectroscopy. **Biochimica et Biophysica Acta**, Amsterdam, v. 975, p. 384394, 1989.

POTT, A.; POTT, V. J. **Plantas do Pantanal.** Brasilia, DF: Embrapa Informaçao Tecnològica, Corumbà: Embrapa Pantanal, 1994. 320 p.

RATTER, J. A.; RICHARDS, P. W.; ARGENT, G.; GIFFORD, D.R. Observations on the forest of some mesotrophic soils in Central Brazil. **Revista Braseira de Botânica**, Sao Paulo, v.1 p. 47-58, 1978.

RÊGO, G. M.; POSSAMAI, E. **Evaluation of chlorophyll levels in the growth of jequitibà-rosa (*Cariniana legalis*) seedlings**. Colombo: EMBRAPA, 2004 (Technical Communication, 128).

REINHARDT, F. **Down to earth, applying bisiness principles to environmental management**. Harvard Business School Press. Cambridge. 2000.

REITZ, R.; KLEIN, R. M. & REIS, A. **Projeto Madeira de Santa Catarina**. Itajai: Herbàrio Barbosa Rodrigues, 1978. 320 p.

REY, P.; PRUVOT, G.; GILLET, B.; PELTIER, G. Molecular chracterization of two chloroplastic proteins induced by water deficit in *Solanum tuberosum* L. plants: involvement in the response to oxidative stress. In: Smallwood, M.F.; Calvert, C.M.; Bowles, D.J. (eds.).

Plant responses to environmental stress, p. 145-152. Oxford, BIOS Scientific Publishers, 1999.

RIBEIRO, R. V.; MACHADO, E. C. Some aspects of citrus ecophysiology in subtropical climates:

re-visiting photosynthesis under natural conditions. **Brazilian Journal Plant Physiology**, Londrina, v. 19, n. 4, p. 393-411, 2007.

ROCHA, A. M. S.; MORAES, J. A. P. V. Influence of water stress on gas exchange in potted young plants of *Stryphnodendron adstringens* (mart.) coville. **Revista Brasileira de Fisiologia Vegetal**, v.9, n.1, p. 41-46, 1997.

RODRIGUES, B. M.; ARCOVERDE, G. B.; ANTONINO, A. C. D.; SANTOS, M. G. Water relations in physic nut according to climatic seasonality, in semiarid conditions. **Pesquisa Agropecuària Brasileira**, Brasilia, v.46, n.9, p. 1112-1115, 2011.

ROZA, Francisvaldo Amaral. **Morphophysiological changes and water use efficiency in *Jatropha curcas* L. plants subjected to water deficiency.** 2010. 67 f Dissertation (Master's Degree in Plant Production) - State University of Santa Cruz, Ilhéus-Ba, 2010.

SALOMÂO , A. N.; ALLEM, A. C. Polyembryony in angiospermous trees of the Brazilian cerrado and caatinga vegetation. **Acta Botanica Brasilica**, v. 15, n. 3, p. 369-378, 2001.

SAUSEN, T. L. **Physiological responses of *Ricinus communis* to reduced soil water availability**. 2007. 71p. Master's dissertation in Plant Science - Federal University of Rio Grande do Sul, Rio Grande do Sul - Porto Alegre - RS, 2007.

SCALON, S. P. Q.; MUSSURY, R. M.; EUZÈBIO, V. L. M.; KODAMA, F. M.; KISSMANN, C. Hydric stress on the metabolism and initial growth of Mutambo seedlings (*Guazuma ulmifolia Lam.*) **Ciência Florestal**, Santa Maria, v. 21, n. 4, p. 655-662, 2011.

SCIPIÂO, Tatiana Teófilo. **Industrial policy to promote local productive arrangements: a case study in Marco, Ceará**. 2004. Dissertation (Master's Degree in Industrial

Pùblicas e Sociedade) - Center for Applied Social Studies, State University of Cearà, Fortaleza, 2004.

SEBRAE. Brazilian Micro and Small Business Support Service. **Development Plan for the Marco Furniture Local Productive Arrangement - Cearâ.** Fortaleza, 2008. 158 p.

SILVA, E, C.; NOGUEIRA, R. J. M. C.; AZEVEDO NETO, A. D.; BRITO, J. Z.; CABRAL E. L. Ecophysiological aspects of ten species in a caatinga area in the municipality of Cabaceiras, Paraiba, Brazil. **Iheringia**, Sér. Bot., Porto Alegre, v. 59, n. 2, p. 201-205, 2004.

SILVA, E. C. *et al.* Stomatal changes induced by intermittent drought in four umbu tree genotypes. Brazilian **Journal of Plant Physiology**, Londrina, v. 21, n. 1, p. 33-42, 2009.

SILVA, E.C.; NOGUEIRA, R.J.M.C.; AZEVEDO NETO, A.D.; SANTOS, V. F. Stomatal behavior and leaf water potential in three woody species grown under water stress. **Acta Botanica Brasilica**, v.17, p.231-246, 2003.

SILVA, M. A. V.; NOGUEIRA, R. J. M. C.; OLIVEIRA, A. F. M.; SANTOS, V. F. Stomatal response and dry matter production in young mastic plants submitted to different water regimes. **Revista Arvore**, Viçosa, v.32, n.2, p.335-344, 2008.

SILVA, S.; SOARES, A. M.; OLIVEIRA, L. E. M. Physiological responses of promising grasses for riparian revegetation of hydroelectric reservoirs, submitted to water deficiency. **Ciência e Agrotecnologia**, v. 25, n. 1, p. 124-133, 2001.

SOUZA, S.M.; LIMA, P.C.F. **Characterization of seeds of some native forest species of the Northeast**. In: CONGRESSO NACIONAL SOBRE ESSÊNCIAS NATIVAS, 1982, Campos do Jordao. Proceedings. Sao Paulo: Forestry Institute, 1982. p.1156-1167. Published in Silvicultura em Sao Paulo, 1982, v.16 A, part 2, 1982.

STAPE, J.L., BINKLEY, D., RYAN, M.G. Eucalyptus production and the supply, use and efficiency of use of water, light and nitrogen across a geographic gradient in Brazil. **Forest Ecology and Management**, v.193, p.17-31. 2004.

TAIZ, L.; ZEIGER, E. **Plant physiology**. 5. ed. Porto Alegre: Artmed, 2009. 918p.

TATAGIBA, S. D; PEZZOPANE, J. E. M.; REIS, E. F. Water relations and gas exchange in the early selection of eucalyptus clones for environments with different soil water availability. **Floresta**, Curitiba, v. 38, n. 2, p.387-400, 2008.

TEREZO, E. F. de M. **Status of mahogany (Swietenia macrophilla, King) in the Brazilian Amazon**. Brasilia: Ministry of the Environment, 2002. 47 p.

TOMZAC, Volmar Elias. **Ecophysiology of ipê-roxo seedlings (Tabebuia impetiginosa Mart. ex. DC. Standal.) subjected to water stress**. 2012. 63 f. Dissertation (Master's Degree in Plant Science) - Universidade Federal Rural do Semi-Arido, Mossoró-RN, 2012.

TONELLO, K. C.; FILHO, J. T. Ecophysiology of three native tree species from the Brazilian Atlantic Forest under different water regimes. **Irriga**, Botucatu, v.17, n.1. p. 58-101, 2012.

TORRES NETTO, A. *et al.* Photosynthetic pigments, nitrogen, chlorophyll a fluorescence and SPAD-502 readings in coffee leaves. **Scientia Horticulturae**, v. 104, n. 02, p.199-209, 2005.

TROVÂO, D. M. B. M.; FERNANDES, P. D.; ANDRADE L. A.; NETO J. D. Seasonal variations in physiological aspects of Caatinga species. **R. Bras. Eng. Agric. Ambiental**, v.11, n.3, p.307-311, 2007.

VALENTE, O.F.; GOMES, M.A. **Conservação de nascentes: hidrologia e manejo de bacias hidrogràficas de cabeceiras**. Viçosa-MG: Aprenda Fàcil, 2005. 182 p.

VAN DIJK, A. I. J. M.; KEENAN, R. Planted forests and water in perspective. **Forest Ecology and**

Management, v. 251, p. 1-9, 2007.

VICTÓRIO, C. P.; KUSTER, R. M.; LAGE, C. L. S. Light quality and photosynthetic pigment production in in vitro plants of *Phyllanthus tenellus Roxb*. **Revista Brasileira de Biociências**, Porto Alegre, v. 5, p. 213-215, jul. 2007.

VITAL, A.R.T., LIMA, W.P., CAMARGO, F.R.A. Effects of clear-cutting a Eucalyptus plantation on the water balance, water quality and soil and nutrient losses in a watershed in Vale do Paraiba, SP. **Scientia Forestalis**, v.55, p.5-16. 1999.

WARREN, C.R.; ARANDA, I.; CANO, F.J. Responses to water stress of gas exchange and metabolites in Eucalyptus and Acacia spp. **Plant, Cell and Environment**, v.34, p.1609–1629, 2011.

WOODCOCK, D.Wood specific gravity of trees and forest types in the southern Peruvian Amazon. **Acta Amazonica**, Manaus v.4, n.3, p.589-599, 2000.

ZENID, G. J. **Wood for furniture and construction**. Department of Science, Technology and Economic Development of the State of São Paulo, 2000. 1 CD-ROM. (IPT. Publication 2779).

Printed by Books on Demand GmbH, Norderstedt / Germany